Optics Experiments and Demonstrations for Student Laboratories

IOP Series in Emerging Technologies in Optics and Photonics

Series Editor

R Barry Johnson a Senior Research Professor at Alabama A&M University, has been involved for over 50 years in lens design, optical systems design, electro-optical systems engineering, and photonics. He has been a faculty member at three academic institutions engaged in optics education and research, employed by a number of companies, and provided consulting services.

Dr Johnson is an IOP Fellow, SPIE Fellow and Life Member, OSA Fellow, and was the 1987 President of SPIE. He serves on the editorial board of Infrared Physics & Technology and Advances in Optical Technologies. Dr Johnson has been awarded many patents, has published numerous papers and several books and book chapters, and was awarded the 2012 OSA/SPIE Joseph W Goodman Book Writing Award for Lens Design Fundamentals, Second Edition. He is a perennial co-chair of the annual SPIE Current Developments in Lens Design and Optical Engineering Conference.

Foreword

Until the 1960s, the field of optics was primarily concentrated in the classical areas of photography, cameras, binoculars, telescopes, spectrometers, colorimeters, radio-meters, etc. In the late 1960s, optics began to blossom with the advent of new types of infrared detectors, liquid crystal displays (LCD), light emitting diodes (LED), charge coupled devices (CCD), lasers, holography, fiber optics, new optical materials, advances in optical and mechanical fabrication, new optical design programs, and many more technologies. With the development of the LED, LCD, CCD and other electo-optical devices, the term 'photonics' came into vogue in the 1980s to describe the science of using light in development of new technologies and the performance of a myriad of applications. Today, optics and photonics are truly pervasive throughout society and new technologies are continuing to emerge. The objective of this series is to provide students, researchers, and those who enjoy self-teaching with a wide-ranging collection of books that each focus on a relevant topic in technologies and application of optics and photonics. These books will provide knowledge to prepare the reader to be better able to participate in these exciting areas now and in the future. The title of this series is Emerging Technologies in Optics and Photonics where 'emerging' is taken to mean 'coming into existence,' 'coming into maturity,' and 'coming into prominence.' IOP Publishing and I hope that you find this Series of significant value to you and your career.

Optics Experiments and Demonstrations for Student Laboratories

Stephen G Lipson

Emeritus Professor of Physics and Electro-Optics, Technion-Israel Institute of Technology, Haifa, Israel, and Professor of Physics and Optical Engineering at Ort Braude College of Engineering, Karmiel, Israel

IOP Publishing, Bristol, UK

Multimedia content is available for this book from http://iopscience.iop.org/book/978-0-7503-2300-0.

ISBN   978-0-7503-2300-0 (ebook)
ISBN   978-0-7503-2298-0 (print)
ISBN   978-0-7503-2301-7 (myPrint)
ISBN   978-0-7503-2299-7 (mobi)

DOI   10.1088/978-0-7503-2300-0

Version: 20200801

IOP ebooks

British Library Cataloguing-in-Publication Data: A catalogue record for this book is available from the British Library.

Published by IOP Publishing, wholly owned by The Institute of Physics, London

IOP Publishing, Temple Circus, Temple Way, Bristol, BS1 6HG, UK

US Office: IOP Publishing, Inc., 190 North Independence Mall West, Suite 601, Philadelphia, PA 19106, USA

This book is dedicated to my dear wife, Rina Lipson, and also to the memory of my father, Henry Lipson FRS (1910–91), who introduced me to practical optics when I was very young.

Contents

Preface

During the 50 years that I have been teaching optics, I have made a point of combining theory lectures with experimental laboratory work. Of course, it isn't possible to let students discover for themselves every facet of the subject in the teaching labs, but optics is without doubt one of the disciplines where a large fraction of the subject is accessible to amateur scientists without a great need for technical assistance. Asking students later in life what they remember best from their university studies, it is usually related to the labs and to lecture demonstrations (particularly those that went wrong in front of a whole class!). I have been involved in the development of teaching laboratories in several institutions, both in general physics and particularly in my specialty of classical optics. I hope that the compendium of ideas which I want to present in this book will be an inspiration to others with similar tasks. Most of the experiments derive from my experience in the third-year and Advanced Physics labs at Technion, in the Geometrical Optics, Physical Optics and Advanced Optical Engineering labs at Ort-Braude Engineering College, Karmiel, in the Modern Optics lab at Cornell University, and from visits to other universities.

This book evolved from what was originally intended to be a chapter in the 4th edition of the book *Optical Physics* (Cambridge University Press) which I co-authored. However, it quickly turned out that much more than a chapter was needed to do justice to the topic, and a separate publication was needed.

I know that the scope of the experiments described in the text is limited by my practical experience and does not cover the field completely in any way. Obviously several important topics, such as fibre optics, quantum and computational optics are hardly mentioned. As a result, I would be happy to hear from readers with suggestions about additional experiments which they have devised and have been carried out at the student level without specialized equipment. There would maybe be sufficient interest that I could organize them into a website which could be continuously updated and brought to the notice of interested teachers. I can be contacted at sglipson@ph.technion.ac.il.

I should like to thank the staff of IOP Publishing, particularly Ashley Gasque, Robert Trevelyan and Heather McKenna, for their prompt and dedicated assistance in producing this book.

Stephen G Lipson, Haifa, July 2020

Acknowledgements

I am grateful to many of my colleagues and students for useful ideas and input about many of the experiments. I would like to mention in particular Jan Gennosar, who has commented on much of the manuscript, and Rami Aharoni, Eric Akkermans, Gidi ben Yosef, Michael Berry, Eberhard Bodenschatz, David Gershoni, Modi Hirshhorn, Ariel Lipson, Said Mahajana, Amnon Nakar, Eli Raz, Michael Revzen, Erez Ribak, Joel Seligson, Eyal Schwartz and Atef Shalabney, who have worked with me for many years, introduced me to several of the experiments described here, and helped me to solve practical problems. In addition, I am grateful to the high-school, undergraduate and graduate students, who have developed and performed many of the experiments described in the book: Jonathan Amasay (section 4.6), Ananda Atalla (sections 4.4, 8.1), Alexandra Bakman (section 4.2), Anan Basis (section 7.4), Tomer ben Aroush (section 8.3), Sheli ben Naim (sections 3.3, 6.1), Saber Boulahjar (section 8.3), Karin Bronstein (section 6.1), Tal Dahan (section 4.1), Assaf Diringer (section 5.3), Natalie Ettingen (section 6.1), Amir Feigenbaum (sections 2.7, 7.4) Sarit Feldman (sections 4.4, 5.4), Daniel Gal (sections 3.3, 6.1), Yarden Gal (sections 3.3, 6.1), Guy Golan (sections 5.6, 7.6), Anatoly Goldman (section 7.4), Alon Granek (section 5.2), Anton Grodeski (section 4.2), Eman Hasoun (section 3.3), Shahar Hirshfeld (sections 5.4, 6.7), Lobna Khanger (sections 4.4, 5.2), Inbal Kom (sections 7.4, 7.5), Eli Levy (section 4.5), Michal Lipson (section 4.5), Marwa Mahajna (section 3.3), Nir Manor (section 4.1), Ilay Mengel (section 5.5), Itai Mengel (section 7.6), Dvir Mizrachi (sections 3.3, 6.4), Ariel Notcovitch (section 7.5), Tal Obstbaum (sections 7.4, 7.5), Taher Odeh (sections 4.2, 8.1), Osnat Ostfeld (section 7.5), Evgeny Ostrovsky (section 4.7), Giora Peniakov (section 10.1), Boaz Ran (section 7.5), Tav Rotenberg (section 2.9), Carmel Rotschild (section 4.5), Ron Ruimy (section 7.5), Ilya Sabenko (section 7.4), Gal Shenkar (section 2.9), Yuval Shapira (section 6.4), Gal Tois (section 4.1), Omer Tores (sections 5.6, 7.6), Chen Turgeman (sections 4.4, 5.2), Roman Vander (sections 6.7, 7.5), Elroi Vinderboim (sections 3.3, 6.4), Eve Zisselman (section 4.7).

Author biography

Stephen G Lipson

Stephen G Lipson is Emeritus Professor of Physics and Electro-optics at Technion and Ort Braude College in Israel. He received his PhD from the University of Cambridge and has been at Technion since 1966. His research has centred around using optical methods in condensed matter physics and in 1973 he contributed to the initial development of Adaptive Optical systems for astronomical telescopes while working for Itek Corporation. He is a coauthor of the textbooks *Optical Physics* (2011, 4th ed, Cambridge University Press) and *An Introduction to Optical Stellar Interferometry* (2006, Cambridge University Press) and has received awards for academic-industrial research cooperation.

IOP Publishing

Optics Experiments and Demonstrations for Student Laboratories

Stephen G Lipson

Chapter 1

Introduction

1.1 What is the purpose of this book, and for whom it is intended

The optics experiments described in the book are all[1] home-designed and can be set up by students themselves using commonly-available equipment. I have personally carried out all of them, or supervised their performance by students. They are intended to illustrate the basic physics of their subjects, as well as their applications outside the laboratory in some instances. I give a brief introduction, with references to the underlying theory, try to describe the basic plan and purpose of each experiment and in particular to emphasize points which can make the difference between a successful experiment and a failure. In many cases I present sample data as a background to discussing particular ideas. I am not trying in any way to provide laboratory manuals for students to use directly, because each institution has its own style of laboratory teaching, and this is of course dependent on factors such as the lab time, space and equipment available. Some of the experiments can be used as lecture or lab demonstrations. Some relate to basic material which every student might be required to carry out in a single lab session[2]; some would be in a list of options[3], from which a student can choose one or two; some are illustrative of an advanced topic which is related to a research field, and might be offered for choice in an advanced or project laboratory[4], with a whole semester for completion. In most cases the experiments should be carried by students in pairs; it is always good to hear a different opinion about how the experiment should be done, and each one learns

---

[1] Except section 10.1, on photon correlations.

[2] For example: Paraxial imaging by a lens. Here, correction for the thickness of the lens must be included in order to get good results.

[3] For example: Holography; the idea is classical, but its implementation using a digital camera introduces new considerations, limitations and possibilities.

[4] For example: Surface Plasmon Resonance, which is a widely used industrial technique.

doi:10.1088/978-0-7503-2300-0ch1

from the other, but if there are three working on it, often one becomes a bystander while the others do the work.

Several experiments of relevance are available commercially as kits. In general, these experiments minimize the experience and pleasure of setting up an experiment oneself and getting it to work. However, I must admit that some of them introduce students to methods and techniques which are beyond the average student's capabilities, but which are of great educational value. I will not refer to these experiments in the book.

1.2 Basic equipment: hardware, light sources, lenses, mirrors, windows, filters, cameras etc

1.2.1 Standard equipment

One aim of this text is to suggest experiments which teach optical principles without needing very specialized equipment. The basic equipment covering most of the experiments includes the following.

1.2.1.1 Optical tables
Tables of size around 1 m × 1.5 m with a square array of tapped holes drilled in them are expensive, but a worthwhile investment. But most of the experiments do not need flotation mechanisms to avoid vibration. When working on a table of this size and larger, it is a good idea to avoid backache by arranging the optical system near the edges, so that the elements needing adjustment are easy to reach. For the few interference experiments where vibration isolation is necessary, a small plate with tapped holes standing on a partially-inflated tyre inner-tube should fit the bill.

1.2.1.2 Vertical blocking plates
These are simply vertical metal or plastic plates painted matt black which block light beams so that they do not interfere with other experiments in the lab. They also can be used to prevent extraneous light getting into an experiment. They should be designed to have an L shape so that they can easily be bolted to the optical table.

1.2.1.3 Opto-mechanical hardware
Some researchers prefer to attach their optical elements mounted on rods directly to the tables using post-holders; my opinion is that this leaves too many degrees of freedom for beginning students and the use of optical benches (rails) for student experiments is preferable in most cases. Optical benches of several lengths are needed. They can easily be connected in series to make longer units; one should remember that a carrier may have to cross a joint continuously and most commercial rails are designed for this possibility. The rails should have length scales attached to them, which are important for several of the experiments. If rails are connected in series, be prepared for errors that may be introduced when the scale pointer moves from one rail to the next! It is a pity if students spend much time looking for the hardware needed to set up their experiments, so a copious and well-organized supply of carriers is advised; some of them should be long enough to be

mounted by two or more optical elements which then move together (e.g. the telephoto lens combination in section 2.4). Many mounting rods of different lengths should also be available; again, short ones can be joined together. And a good supply of screws of different types; please try not to make the mistake of having more than two or three common screw threads in the lab (e.g. M6 and M4 only), and certainly not threads of similar diameters which can be mixed up, such as 1/4 inch and M6. Lens and mirror mounts up to 50 mm diameter with tilting mechanisms are needed, although most experiments in this book do not need accurate tilt adjustment for lenses. Professional holders for vertical samples and slides and some horizontal samples, as well as prism and cube-beam-splitter mounts are convenient, but often can be improvised. Several of the experiments require micrometer translation stages, which are quite expensive; a resolution of 10 μm is usually sufficient, although adjusting fringes in an interferometer may need better sensitivity. Micrometer rotation stages are also needed in some experiments but coarser rotation stages are sufficient for most of the experiments here. Finally, I recommend spatial filter alignment mechanisms using a microscope objective and pinhole for expanding and cleaning laser beams (section 1.2.2 and figure 1.2).

1.2.1.4 Optical elements

A good supply of lenses and other components is necessary; again, one does not want students to waste too much time hunting for commonly used components. Commercial sets of singlet lenses, 25 mm diameter, are suggested; the lenses should be mounted in robust labelled mounts *before* they fall on the floor. A certain number of experiments require achromats, and a selection should be available; likewise, several microscope objectives of various magnifications. A few spherical mirrors and larger lenses are good to have, although not an integral part of the experiments here. Front-surfaced plane mirrors 25 mm and 50 mm diameter, of $\lambda/4$ quality, are needed for many experiments. Then prisms are required, both 60° and 90°. A rough calculation shows that a spectrometer with 25 mm optics requires a prism with about 40 mm sides, because of the oblique incidence. To this list we can add colour filters (narrow and broad band), polarizers ('Polaroid' film is usually sufficient, but higher-quality polarizers, supplied for cameras, are easily available[5]), neutral density filters, optical windows of $\lambda/4$ quality. Beam-splitters are a necessity; it is worth investing in 38 mm non-polarizing cubes to make alignment easier, though some experiments can use plate beam-splitters, which are considerably cheaper.

1.2.1.5 Light sources

Most laboratories traditionally use Ne–Ne lasers, although recently some are going over to laser diodes. The latter are cheaper, but it is more difficult to get most of the light into a clean parallel Gaussian beam from them. For holography, it is useful to have 10 mW lasers, but for most other experiments 1–5 mW is sufficient. Getting a

---

[5] Be aware that some photographic polarizers have a quarter-wave-plate attached to them. This need not be a problem if you know which side the wave plate is attached; you put the wave-plate on the side where the wave is unpolarized.

large coherent beam requires a spatial filter, with microscope objective and pinhole, as discussed below.

In addition to lasers, broad-band sources are also required. As a general source of white light, battery-operated LED flashlights are very convenient and cheap. Also, spectral sources, particularly Hg and Na discharge lamps are needed for some experiments. One should remember that these sources can not be focused or concentrated since they are spatially incoherent, so the best way to get as much light as possible into a slit or aperture or fibre is simply to put the lamp as close to it as possible! Rarely, a condenser lens is needed—for example if a very large angular distribution of light from a slit is required.

1.2.1.6 Electro-optical equipment

Two instruments which are necessary in many experiments are power meters and cameras. On-line cameras are an integral part of many experiments (e.g. diffraction) and a frame size of 2500×2000 pixels is sufficient for most experiments. Colour is an advantage, but a lot of good work can be done with monochrome cameras, which usually have more pixels. On the other hand, for holography, very high resolution and pixel count are needed, and a professional reflex camera is recommended. Indeed, the colour here is unnecessary, but maybe the camera allows this to be traded for a higher pixel count. Cameras and detectors usually need to be calibrated; a simple procedure is discussed below.

1.2.1.7 Spectrometers

Experiments on dispersion by prisms and diffraction gratings are conveniently done with a spectrometer. Reading the angle scales on conventional spectrometers, to an accuracy of a few minutes of arc by means of a Vernier, is quite difficult for students, and I have seen models with electronic readout with this accuracy by means of an encoder.

1.2.2 Common procedures: alignment of components, cleaning optics, spatial filtering a laser beam, calibrating a camera or detector

1.2.2.1 Aligning lenses along an optical bench

Many experiments need several lenses aligned along an optical bench, and we will describe here a method of getting them concentric, to a reasonable degree of accuracy. First, a weak laser beam (1 mW is sufficient) is arranged to define the optical axis, which should be parallel to the bench at a convenient height for the experiment. At the end of the bench, the beam should strike a vertical fixed card on which its position is marked; if this card is mounted on a carrier, it can be slid along the bench to ensure that the beam height (y) and transverse position (x) are indeed independent of z. The lenses are then mounted on the bench successively, starting with the one most distant from the laser. Its (x,y) position can be adjusted to maintain the centricity of the laser beam around the marked point. Following this, its tilt (around y and x) can be adjusted by ensuring that the beams reflected from the lens go back to the laser, that the reflections from the two surfaces coincide and are symmetrical around the beam exit (it may be necessary to block the beam transmitted by the lens for this stage, in order to avoid interference with beams reflected by other lenses). Note that the tilt around x is not available for adjustment on many mounts. The lenses do not need to be in their final z-positions for alignment if the axis is parallel to the rail, and it is a good idea to check that the alignment does not change as they are slid along the bench. In principle, each lens could be adjusted independently alone on the bench by this method, before the final arrangement is assembled, but small readjustments would still be necessary.

1.2.2.2 Cleaning optics

Optical elements should be kept clean, first by ensuring that they are always held by their edges or inactive surfaces, and second by cleaning off fingerprints as soon as possible after they appear. For the latter purpose, ethanol is best, and at the level of student laboratories lenses and windows can be cleaned by using a small amount of the solvent on lens-cleaning tissue, followed by a dry wipe. The lens may need to be removed from its mount for this procedure to clean it out to its edges.

Mirrors and other coated surfaces need more careful treatment. One way is to lay the mirror on the table, face upwards, with a strip of card or other obstruction attached to the table next to it as a ridge to prevent the mirror sliding. A prism might need to be held in a clip, with the active surface horizontal. Next, place a piece of lens tissue to cover the mirror, and on that put a drop or two of alcohol which will quickly spread out and contact the tissue to the mirror by capillary forces. Next, wait a short while for the solvent to dissolve the dirt, but not long enough for it to evaporate, and then pull the tissue slowly along the x-direction (figure 1.1), at such a rate that it leaves a dry surface behind it with no drops remaining on the mirror, so that all the dirt stays in the solvent absorbed in the tissue. One or two such passes can clean a mirror surface most thoroughly, without damaging the coatings. It is worthwhile practicing this procedure on a piece of glass before first using it on a mirror. This procedure can be used with other solvents, but their rate of evaporation may make it more difficult.

1.2.2.3 Spatially filtering and expanding a laser beam

Before starting diffraction, interference or imaging experiments using a laser, one usually needs to expand the laser beam, which is basically a Gaussian beam with diameter about 1 mm, into a similar beam of diameter 1–3 cm. Moreover, the beam as emitted by the laser contains speckle and other spatial noise, which needs to be eliminated. This is done by using a standard spatial filter (figure 1.2), which consists

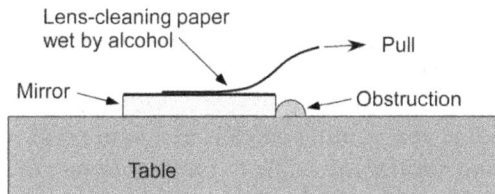

Figure 1.1. Procedure for cleaning a mirror.

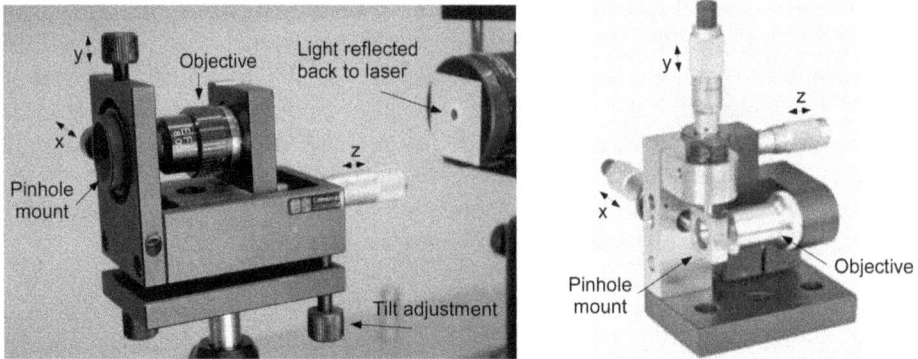

Figure 1.2. Two common versions of the spatial filter assembly: left: Edmund Scientific, with a 5 mW He–Ne laser; right: Newport Optics, figure reproduced with permission of Newport Corporation. All rights reserved.

of a microscope objective (10–40×) and a pinhole whose position can be adjusted in three dimensions. The optimum size of the pinhole is about 1.5 times the Airy disc diameter of the focal spot, which depends on the incident laser beam diameter, the focal length of the objective and the wavelength; for a 20× objective it is in the region of 10–15 μm diameter. The laser beam has to be focused exactly onto the pinhole, after which most of the noise is eliminated from the emerging beam and it can be collimated by a converging lens, whose focal length determines the output beam diameter. The most trying part of the setup is to adjust the filter so that it produces a clean Gaussian beam, concentric with the original laser beam, which in many cases has already been used before expansion to align the various optical elements in the experiment.

A procedure which has been found to work satisfactorily is as follows: (see also the movie and figure 1.3).

1. Unless the laser is weak (5 mW or less), it is recommended to insert a neutral density (ND) filter in order to reduce the beam to that level, so that the procedure can be carried out without using eye protection spectacles.

2. The original direction of the laser beam (horizontal, and parallel to the optical bench if one is used) is marked on a screen distant 30–40 cm from the laser. The spatial filter system is then mounted at a distance of about 10 cm from the laser, with its axis at the right height and lateral position and mechanically parallel to the direction of the beam (z), as judged by eye.

3. If the pinhole is supported by magnets, or is screwed into a holder and can easily be taken out without disturbing the filter assembly, the exact position and alignment of the spatial filter objective lens can be adjusted to provide the required output beam by first checking that the expanding wave is centred around the laser beam mark on the screen. The pinhole can then be returned to its position, and continue with stage 5. Otherwise, continue with stage 4.

4. However, in some spatial filter models extracting and replacing the pinhole are not easy, and springs are liable to jump out and get lost! In this case, the height and lateral position of the system are next adjusted more accurately by requiring the light beam reflected back from the objective to create a

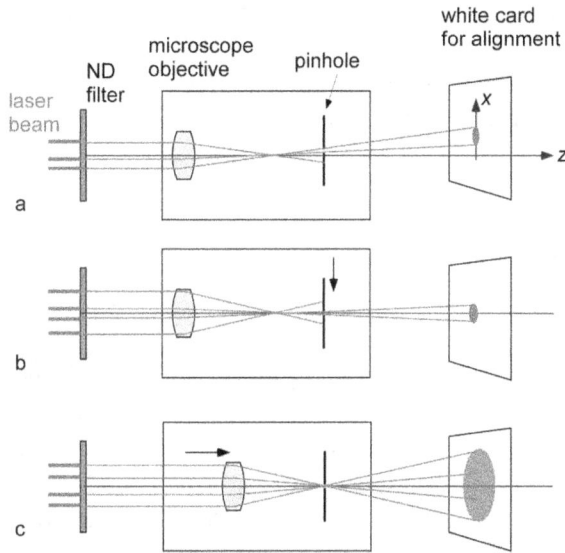

Figure 1.3. Alignment steps. A movie is available online at http://iopscience.iop.org/book/978-0-7503-2300-0.

symmetrical patch surrounding the exit from the laser. The reflected beam is diverging, so that at the distance of a few cm the patch is quite large and this can be done quite accurately. If necessary, a white paper slip with a small hole in its centre can be attached to the laser or ND filter, with the beam going through the hole, in order to see the reflected beam better (see figure 1.2(left)).

5. Using the z-axis (micrometer) screw, the pinhole is brought to a position along the z-axis which is considerably further from the objective than its focal length (figure 1.3(a)). Using a white card (a visiting card is a convenient size) close to the pinhole, one looks for some weak light coming through the pinhole. This light intensity is then maximized by playing with the x–y controls on the pinhole. As a result, the pinhole is now close to the axis (figure 1.3(b)), and a light patch can be seen on the card.

6. Using the z-axis screw, the pinhole is brought closer to the focal point; the intensity of the light patch grows at first, but when it starts to fall, the x–y adjustments can be used to bring the intensity back. At first, these adjustments are not sensitive, but as the pinhole approaches the focal point they become more and more critical and diffraction rings can be seen in the patch. Finally, when the patch is very bright and the first diffraction ring has essentially disappeared, the filter is in its optimal position and any further change in x, y or z only spoils it (figure 1.3(c)).

7. Without a collimating lens, the beam patch on the screen or wall should now surround the mark of the original laser beam quite symmetrically; if not, small adjustments to the height and lateral position of the filter assembly can be made to correct this, the x–y of the pinhole being continuously adjusted to keep the intensity maximum. Confirm that the beam reflected back into the

laser remains symmetrical (stage 4); this may require small angular adjustment to the assembly.

8. Finally, a collimating lens is inserted and its position used to bring the expanded beam into alignment with the original laser beam.

1.2.2.4 Calibrating a camera or detector

To calibrate a camera or detector we need a source of controllable intensity. This is best provided by an incoherent lamp operated by a stabilized power supply, followed by a pair of polarizers whose relative orientation can be varied and measured. One should not use a laser for calibrating a camera, because the received signal is the average over a speckle pattern, and this is likely to contain spots which saturate some pixels, even if the average is within the dynamic range. There is no need for very high quality polarizers, provide that they are uniform and clean, so that the only change in intensity of the transmitted light comes from the relative angles between the polarizer and analyzer. If there is a possibility that the detector or camera is polarization sensitive, the analyzer can be followed by a quarter-wave plate which rotates with it; such a combination is used by photographers and can be purchased as a single unit. The detector or camera output is then measured as the relative angle is changed. The measured results can then be fitted to the function $I(\theta) = A \sin^2 \theta + B$ to get an idea of deviations from linearity. The examples in figure 1.4 show the results for a CCD camera (expected to be linear between the dark level and saturation level) and CMOS camera, which has a response closer to that of the eye.

1.2.3 Laser safety

I do not intend here to provide a guide to the safe use of lasers in a student laboratory. All teaching institutions have a safety officer whose job it is to give students an obligatory lesson on laser safety before they start in a laboratory where lasers are used, and it is the lab instructor's job to ensure that the safety requirements are carried out. Most of the experiments described in this book either do not require the use of lasers, or can be carried out with lasers whose power is below the danger limit. Classes 1 and 2 visible continuous-wave lasers (<1 mW) will satisfy the requirements of almost all the experiments. The only exceptions are holography (best done with a 5–10 mW laser) and of course the experiments on non-linear

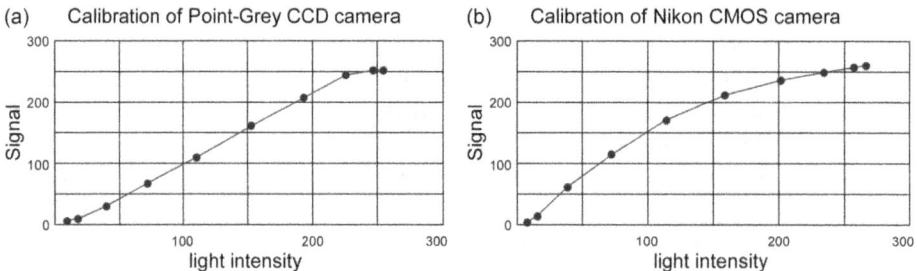

Figure 1.4. Camera calibration: (a) CCD; (b) CMOS. Saturation occurs at 255 grey levels.

optics, where the laser power is an important factor. For these experiments, students must be provided with protective eyewear. This is particularly important when an invisible IR laser is being used (section 7.4).

However, there are some general points about laser safety which are not usually mentioned by safety officers. First, the layout of an optical setup should be planned so that any unexpanded laser beam points in a direction which avoids regions of the lab where people circulate, i.e. they should generally point towards the walls, and black-matt beam-blocking plates (section 1.2.1) should be readily available to stop unwanted beams (reflections, etc). Second, a filter which reduces the intensity should cover the beam exit from the laser while the setup is being aligned, so that accidental reflections or misalignments are not dangerous. Third, high chairs or stools should be provided for seating, so that a laser beam is never close to eye level, either standing or sitting.

IOP Publishing

Optics Experiments and Demonstrations for Student Laboratories

Stephen G Lipson

Chapter 2

Geometrical optics

2.1 Prism spectrometer and glass dispersion

Dispersion, which is the dependence of refractive index on wavelength, is an intrinsic property of any transparent medium. In fact, the only medium that is not dispersive over the whole electromagnetic spectrum is a vacuum! Moreover, the refractive index and the absorption coefficient of a medium are related as functions of light frequency by the Kramers–Kronig relations, which show amongst other things that in a spectral region where the medium is transparent, with a vanishingly small absorption coefficient, the refractive index is always an increasing function of frequency. The dispersion of glass and other bulk solids can be conveniently measured using a spectrometer, which is the basis of this experiment. It requires the material to be uniform in texture and to be prepared in the form of a prism with optically flat plane faces. A useful measure of the dispersion in the visible region is the Abbe dispersion coefficient, defined as $V \equiv \frac{n_y - 1}{n_b - n_r}$, where the subscripts r, b and y refer to red ($\lambda = 656.3$ nm), yellow ($\lambda = 587.6$ nm) and blue ($\lambda = 486.1$ nm) light. These are three prominent lines in the emission spectra of hydrogen and sodium. The dispersion coefficients of glasses are in the range of 70 (least dispersive, i.e. $n_b - n_r$ is minimum) to 20 (most dispersive). Air has $V \approx 90$, water has $V = 72$ and Perspex (PMMA) has $V \approx 50$.

The prism spectrometer is an elementary instrument, which has many applications. It does not provide the wavelength resolution that can be obtained with a diffraction grating spectrometer (section 4.3), but has almost 100% light throughput, which is often important. A grating, on the other hand, divides the light power into several orders of diffraction, including the zero order which has no dispersion. Only with very special blazed gratings (section 4.3.2) can the diffraction efficiency come close to 100% in a particular order, at a particular wavelength.

doi:10.1088/978-0-7503-2300-0ch2

The classical spectrometer, which is suitable for either a prism or a grating, consists of an input slit of variable width which is illuminated by the light source under investigation, followed by a collimator lens, which focuses an image of the slit to infinity. The light beam then passes through the prism, which creates images of the slit in different wavelengths at different angles according to the refractive index dispersion of the prism, and these images are then observed or photographed through a telescope. The observed spectrum, as a function of angle between the telescope axis and the incident beam is then converted to refractive index dispersion. Older spectrometers measure the angles with a mechanical scale and a Vernier to interpolate to an accuracy of about 1/20 degrees or 2 arc-min; more modern versions use an electronic transducer to read out the angles digitally. Before starting the experiment, the optics should be adjusted without a prism; first, the telescope to produce a sharp image of a distant object, and then, looking at the direct transmission direction, the collimator length is adjusted to give a sharp image of the illuminated slit.

When the prism is placed in its position, you may observe that the slit image has moved up or down. In that case, the prism must be levelled, i.e. its apex must be made parallel to the rotation axis, before the measurements are continued. To get accurate results, the slit width must be minimized.

The angle of deviation D is a function of the prism head angle α, the refractive index n and the angle of incidence θ at the prism (figure 2.1). The prism should be rotated to provide minimum deviation; in principle this should be done for each wavelength, but negligible error is introduced by doing it for a wavelength in the centre of the spectrum and fixing the prism position there. Then the refractive index is related to D and α by the relationship

$$n = \frac{\sin[(\alpha + D)/2]}{\sin(\alpha/2)}$$

which should be proved before starting the experiment (hint: use symmetry). Most prisms used for spectrometry have $\alpha = 60°$, but this can be confirmed by rotating the

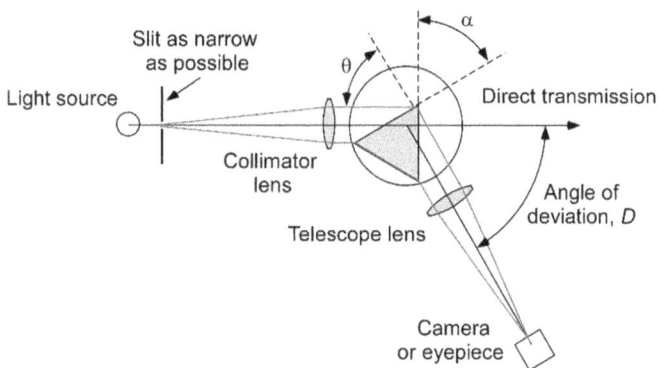

Figure 2.1. Prism spectrometer with camera vision. Note that the optimum position for the prism is not centred on the telescope rotation axis.

prism so that its head points towards the collimator, and beams from two surfaces are reflected; the angle between them (2α) can then be measured using the telescope.

2.1.1 Calibration

The spectrometer must be calibrated using a spectral source with known wavelengths. A low-pressure mercury lamp is ideal for this, since it provides sufficient spectral lines with known wavelengths to fit the refractive index to a curve for interpolation. The relevant lines in the visible spectrum are at 404.6 nm, 435.8 nm, 546.1 nm, 576.9 nm and 579.0 nm. Once the data $n(\lambda)$ is recorded, a suitable curve should be found to fit it within the error bars of the measurements. A polynomial is not good for this, but the Sellmeier equation can be used, although a simpler version of it $n(\lambda) = a + b\lambda^{-2}$ is sufficiently accurate. Following this, the Abbe dispersion coefficient can be found by calculating the three relevant values of $n(\lambda)$ by substitution in this formula, and then the type of glass can be identified using the Schott diagram.

2.1.2 Spectral resolution

In another experiment we will compare prism and diffraction grating spectrometers, so the wavelength resolution should be measured. One way to do this is to see whether the Na and Hg yellow doublets can be resolved. The former has a separation 0.6 nm, the latter 2.0 nm. The slit width is an important parameter here; it should be made as narrow as possible.

2.2 Critical angle of reflection and Abbe refractometer: measurement of refractive index of a fluid

2.2.1 A classroom demonstration of critical reflection at the air–glass interface

The critical angle of reflection at the interface between two optical materials with refractive indices n_2 and n_1 is $\theta_c = a \sin(n_2/n_1)$. A very simple classroom demonstration of critical reflection requires a glass plate (doesn't need to be optically flat) several mm thick, piece of white paper and a laser pointer. The white paper is wetted and placed in contact with the lower side of the glass plate. The laser pointer is set to illuminate a point on the paper (from either side). Light is scattered in the wet paper at all angles, and refracted into the glass in a cone of semi-angle θ_c (water–glass) $\approx 60°$. Light in this cone is reflected back from the top surface of the glass at angles outside a cone of semi-angle θ_c (glass–air), about 40°. So a ring of light with inner angular radius 40° and outer radius 60° is seen reflected onto the paper (figure 2.2).

2.2.2 The Abbe refractometer

The Abbe refractometer was devised by Ernst Abbe in the late 19th century for measuring the refractive index of liquids and some solids. It is still widely used for that purpose. The original instrument was, of course, manually operated, but nowadays digital versions are available although the optical principle has not changed. The accuracy of a refractometer is about 10^{-4} in the visible region.

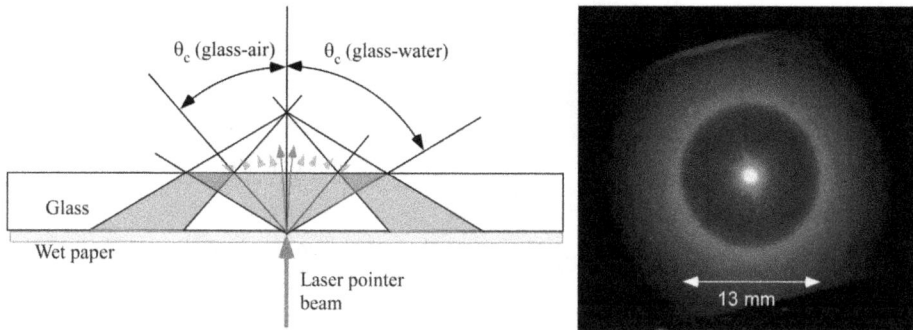

Figure 2.2. Experiment with a laser pointer, a 4 mm thick glass plate and a wet paper scatterer. The inner boundary of the ring of light scattered by the paper is clear; the spot in the centre is a direct image of the illuminating laser spot. The outer boundary of the ring is too weak to be seen. The diameter was determined by repeating the experiment with a wet piece of mm graph-paper. What is the refractive index of the glass?

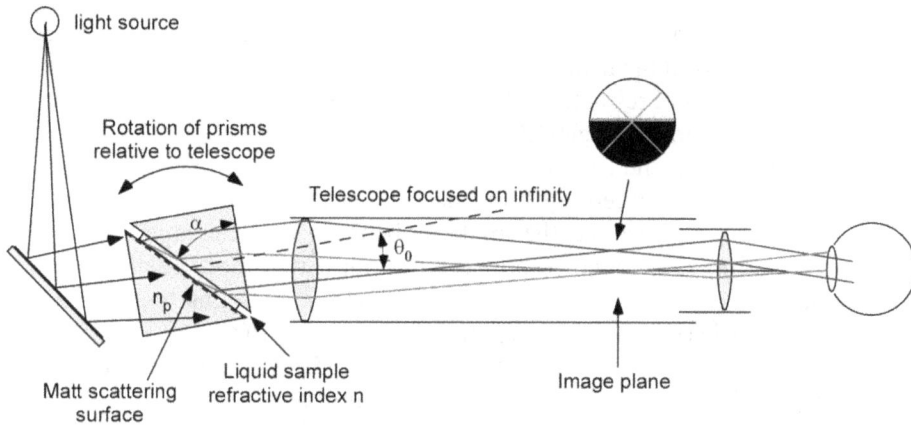

Figure 2.3. Optics of the Abbe refractometer. The rays designated in blue are scattered by the matt surface under the yellow sample at an angle well below the critical angle θ_c, and the red rays marginally below θ_c. The image plane of the telescope indicates angles of rays exiting the second prism. The rays between the prisms are shown in more detail in figure 2.4.

The refractometer allows accurate measurement of the critical angle of reflection at an interface between the sample and the plane surface of glass with known refractive index n_p. A liquid sample is sandwiched between two right-angle prisms, as shown in figure 2.3, and is illuminated by light coming from an extended source, usually the room lights, but maybe a quasi-monochromatic lamp. The light is scattered by the rough matt surface of the first prism, and so it enters the liquid and is incident on the second prism at all angles within a 2π solid angle. Since the second prism has a refractive index greater than that of the liquid, the light exits the second sample-glass surface at all angles of refraction within the prism up to the critical angle $\theta_c = \sin^{-1}(n/n_p)$, which corresponds to 90° incidence within the liquid sample (the red ray in figures 2.3 and 2.4).

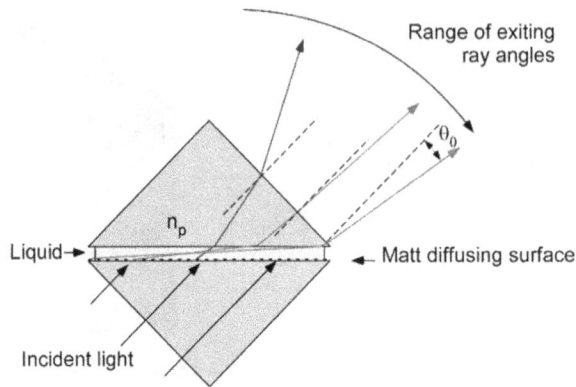

Figure 2.4. Light scattered by the diffusing surface in the liquid is refracted to angles up to the critical angle in the prism. The red ray, incident at 90° to the normal on the upper prism when the liquid thickness is infinitesimal, represents the boundary of rays exiting the prism to the telescope (figure 2.3).

The light leaving the second prism is interrogated by a telescope focused on infinity, so that points in the field of view correspond to angles of exit from the prism and therefore to angles of refraction. In this way, an image is formed which is dark in the region corresponding to angles of refraction above θ_c, and bright below it. When using the refractometer, the user sets the angle of the telescope relative to the prism so that the boundary between light and dark is on the telescope axis, defined by cross-hairs. Thus we determine the angle θ_0 between the telescope axis and the normal to the prism surface, corresponding to critical reflection. Snell's law at the prism–air interface where the light leaves the prism relates this to the refractive indices of the sample (n), the prism (n_p) and the angle α of the prism ($\alpha = 45°$ in the figure) by $\sin\theta_0 = n_p \sin(\alpha - \theta_c)$. Since the refractive indices involved are functions of the wavelength, the boundary between the dark and bright regions has a coloured band along it if the illumination is white, but is sharp if the illumination is monochromatic. A photograph of a manual Abbe refractometer (Hilger and Watts model from about 1950) is shown in figure 2.5.

2.2.3 Using the refractometer to measure the refractive index of a glass plate

The way to measure the refractive index of a parallel-sided plate of glass is to sandwich the plate between the two prisms using a gel or oil with refractive index of value between that of the plate and that of the prism. Most refractometers are supplied with a liquid for this purpose, which has index marginally below that of the prisms. It is quite easy to show using Snell's law that for a system with parallel layers of various refractive indices, the critical angle is determined by the highest index, that of the prism, and the lowest index of the set, which is that of the glass plate.

2.2.4 A lab experiment

A simple lab construction illustrating the idea of the refractometer can be made using a spectrometer to measure the angles. It is illustrated in figure 2.6, the basic

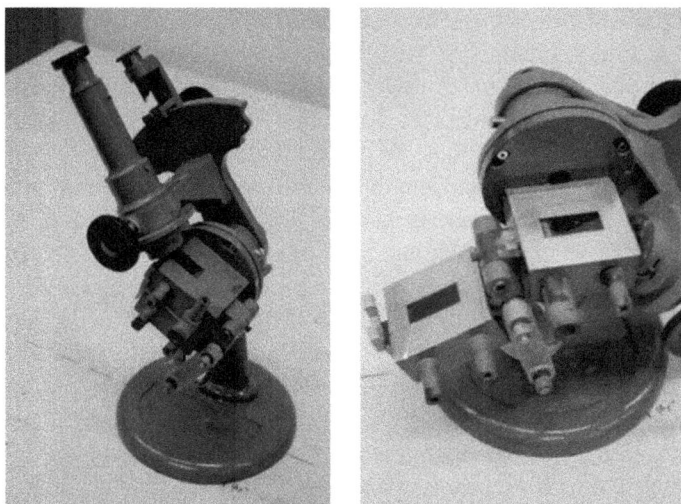

Figure 2.5. A manual Abbe refractometer. The right hand photo shows it opened in order to insert a drop of a fluid sample.

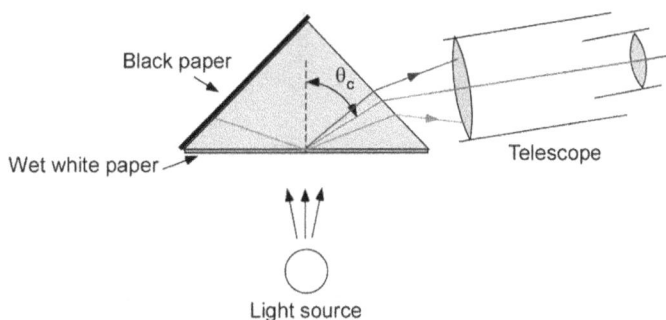

Figure 2.6. A simple experiment which reproduces the idea of the Abbe refractometer. The field of the telescope above the critical angle θ_c (water–glass) sees the black paper; that below θ_c sees the light source scattered by the white paper.

idea being again to use wet paper as the source of diffusely scattered light, and a telescope to measure the maximum angle at which light is refracted within the prism. As in the refractometer, when the telescope is focused on infinity, a half-black field of view is seen, with a tinge of red along the boundary if the source is white.

2.3 Paraxial imaging by singlet lenses: thin lens imaging, Newton's law, depth of field, Scheimpflug construction

This section and the next one are concerned with determining the properties of individual lenses and lens combinations. A simple lens with two surfaces has an axis defined by the line passing through the centres of curvature of its two surfaces; if one surface is a plane, the axis is the normal to it which passes through the centre of curvature of the other one. The most important use of a lens is to create an image, and the way in which it does this can be described in various ways. The most basic lens is a thin lens, and the simplest treatment of this uses paraxial optics, i.e. the propagation of rays which make angles θ with the lens axis small enough for the approximation $\sin \theta = \theta$ to be valid to better than about 1%, which occurs when $|\theta|$ is less than 15°. Then follow the most basic and well-known equations of geometrical optics, which are that for the focal length f in terms of the radii of curvature of the two faces, R_1 and R_2 and the refractive index n (the so-called 'lens-maker's equation'), and the object-image distance equation[1] $1/v - 1/u = 1/f$ or Newton's equation $OF_1 \cdot F_2 I = f^2$. In general, the aims of paraxial geometrical optics are to relate the properties of thick lenses and curved mirror systems as well to these equations, by suitable constructions in which the given optical system is described in terms of a set of cardinal points and planes (section 2.4). We have found that the formulation of ray-tracing in terms of matrices is very convenient for understanding this subject.

By the term 'thin' lens, we mean a lens whose thickness is negligible with respect to its focal length, and therefore its principal planes[2] H_1 and H_2 coincide with the plane of the lens itself. By a 'thick' lens we mean a thick singlet or a coaxial lens combination for which the principal planes H_1 and H_2 do not coincide. Although it might seem that the thin lens is only an ideal concept, in fact there are real systems that fulfill the requirements, such as a glass sphere in air or any lens with two concentric spherical surfaces; in these cases both principal planes pass through the common centre of curvature.

2.3.1 Determination of the focal length of a single converging lens

First, note that the focal planes of the lens can be found using an auto-collimator (section 2.7). For a single lens, the distance between a focal plane and the lens is equal to the focal length, but if the lens is thick, we have to know which principal plane in it represents the equivalent thin lens position for that focal plane. This is the relevant principal plane.

We use elements which can be translated along a long optical rail with a mm scale along it. The elements are mounted on sliders whose positions z can be read on the scale using a cursor, to an accuracy of (say) 0.5 mm. The optical elements are of course each offset from the cursor by a fixed amount, and this has to be taken into account in analyzing the experimental data (constants δ_1 and δ_2 below). All the elements must be adjusted to be at the same height and to be normal to the axis; this can be done visually, although there exist more accurate methods (section 1.2.2). The experiment to determine the effective focal length consists basically of determining a series of object-image pairs. It is convenient to have a fixed object with calibrated dimensions, such as a diffusely back-lit reticle with a scale on it (figure 2.7). Its rail position z_s is constant. The lens is moved along the rail; its rail

---

[1] We use a Cartesian system for axial distances, i.e. distances to the left of the origin are negative, and to the right are positive.

[2] A 'principal plane' (H) of a thick lens or a lens combination is the plane where a single thin lens having the same focal length would be placed in order to image a distant source to the focal plane (F) of the combination. There is a principal plane associated with each ('front' and 'back') focal plane of the system.

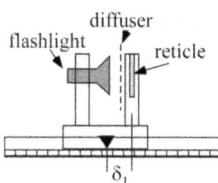

Figure 2.7. Object combination.

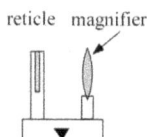

Figure 2.8. Image finder.

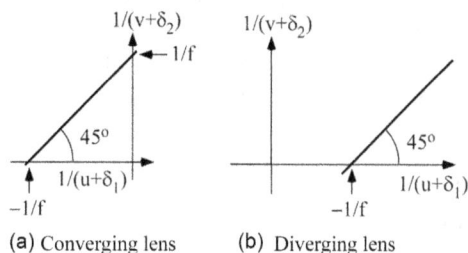

(a) Converging lens   (b) Diverging lens

Figure 2.9. (a) Plot for a converging lens; (b) Plot for a diverging lens.

position is z_L, and the image position z_I is determined. For the latter purpose, it is convenient to use a second transparent reticle with a magnifying glass (about 75 mm focal length) mounted with it on the same slider, and at a distance adjusted for the individual student's convenience (figure 2.8), or maybe a digital camera chip. We will call this the 'image-finder'. This way, the image position can be determined as the point where the image and the reticle are in focus together. The magnification can also be determined. If the nominal focal length of the lens is f, the length of the rail should be at least $7f$ in order to get a range of magnification from 1/5 to 5 (a 1 m rail is suitable for $f < 125$ mm).

As a first stage of the data analysis, one should estimate the offsets between each optical element and the associated rail cursor position, and then calculate the object and image distances ($-u$ and v). Then, using Excel for example, we plot $(1/u)$ against $(1/v)$ and hope to get a straight line satisfying $1/v - 1/u = 1/f$, which has unit slope and intercept $\pm 1/f$ on each axis (figure 2.9). We can now adjust the offsets by adding or subtracting constants δ_1 and δ_2, of order a few mm, to u and v until both the slope of the line and the linear regression fit are both unity (an accuracy of 0.1% can readily be achieved by trial and error). Alternatively, a computer code can be written to optimize δ_1 and δ_2 with the same aim. The magnification should be confirmed to be $(v + \delta_2)/(u + \delta_1)$. It would be interesting to do this experiment on a spherical lens (remember that it needs to be paraxial, which means that an appropriate iris aperture should be associated with it, to admit only rays which are close to the axis) to confirm that this is indeed a 'thin' lens! Certainly, it is very instructive to do the experiment with a meniscus lens.

2.3.2 The focal length of a thin diverging lens

In order to find a (u, v) set for a diverging lens of focal length $f < 0$, one needs a virtual object at a position with positive u, so that the image is real (figure 2.10). This can be obtained by using the same object as before, but augmented by an auxiliary converging lens with focal length about $|f|/2$ to project a real image of the object reticle with about -1 magnification to provide a virtual object. The positions of the auxiliary lens and the virtual object remain constant, and the position of the latter is found with the image-finder used previously. This will also give the scale size of the virtual object, so that the magnification can be determined. Note that since the object and image positions were determined using the same image-finder, $\delta_1 - \delta_2$ is equal to the distance between the principal planes of the lens, which can easily be estimated (about t/n, where t is the axial thickness of the lens and n its refractive index).

2.3.3 The Scheimpflug construction

When there is an extended object which is planar, but its plane is not normal to the optical axis, the paraxial image is also planar and not normal to the axis. The relationship between the two planes for a single thin lens can be found formally by extending the object plane to intersect the plane of the lens, and then constructing the plane defined by the intersection line and the axial image point (figure 2.11). This construction, called the 'Scheimpflug construction', is relevant to landscape and architectural cameras, and was originally derived for the purpose of interpreting aerial photographs taken with balloons. The image is quite distorted (figure 2.12), because the magnification depends on the distance of an image point from the axis. The distortion is either interesting or a nuisance, depending on your point of view! A question: How is the Scheimpflug construction formulated for a thick lens?

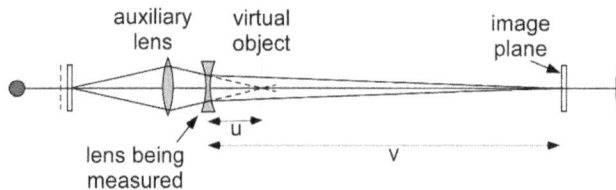

Figure 2.10. Setup for measuring the focal length of a diverging lens.

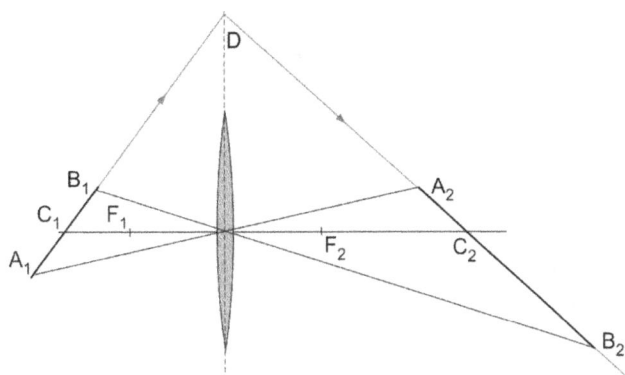

Figure 2.11. The Scheimpflug construction for a thin lens. $A_1C_1B_1$ is the object plane and $A_2C_2B_2$ is the image plane. The ray (in green) $A_1C_1B_1DA_2C_2B_2$ is common to all the object and image points.

Figure 2.12. Sharpest image of a square grid which is situated in a plane at 45° to the optical axis, magnification about 1 at the centre.

2.3.4 Commonly encountered problems

Depth of focus. The image plane is estimated by using the 'no parallax' method, or by judging the sharpness of a camera image. Although this is good for magnifications smaller than unity it becomes less accurate as m increases. An estimate of the variation in estimated image position on repeated trials should show that $\delta(1/v) = \delta v/v^2$ is conserved.

Chromatic aberration. If singlet lenses are used, the values of f and f_{eff} are wavelength-dependent. The telephoto combination magnifies the effect of wavelength, and this could be measured by using several colour filters. To avoid chromatic problems, achromatic doublets can be used for the experiments.

Distortion. If a square grid reticle is used as object, the degree of barrel or pincushion distortion can be observed (section 2.8).

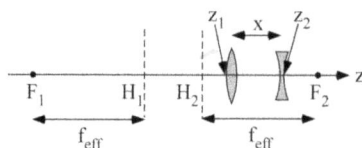

Figure 2.13. Cardinal points of a telephoto combination.

2.4 Compound and thick lenses: focal, principal and nodal planes, zoom lenses

2.4.1 Cardinal points and planes of a compound or thick lens

A thick lens, or a coaxial lens combination, cannot simply be described in terms of its focal length f_{eff}. We therefore consider this experiment more from the point of view of understanding the ideas of cardinal points and planes (principal, H_1 and H_2; focal, F_1 and F_2; nodal, N_1 and N_2; figure 2.13) and effective focal length (f_{eff}) with less emphasis on accuracy in their determination. As mentioned in section 2.3, the principal planes are the planes where an ideal thin lens with focal length f_{eff} would be placed to image an infinitely distant object into the focal plane, one for each focal point. The nodal points are axial points around which the lens can be rotated without causing lateral movement of the image of the distant point; in air ($n = 1$) the nodal and principal points coincide.

2.4.2 Telephoto combination

The telephoto lens combination (basis of the zoom lens[3]) has principal planes at considerable distances from the actual lenses, so it is a good example for an experiment. We have found the pair $f_1 = +75$ mm, $f_2 = -50$ mm, separation $x \cong 50$ mm to be ideal. For this value of x, its effective focal length $f_{\text{eff}} = +150$ mm, and the principal points are on the side of the converging lens at $z = z_1 - 150$ mm, $z = z_2 - 100$ mm, where z_1 and z_2 are the actual lens positions. The two lenses should be attached to a single carrier (hence $x \approx 50$ mm or 2 inch!), which can easily be mounted on the rail in either orientation. A long strip of cardboard, on which to mark the focal and principal plane positions when they are determined, should be mounted on the same carrier, in such a way as not to interfere with other carriers on the same rail. This strip is very useful in understanding the basic ideas.

2.4.3 Determining the focal planes and effective focal length

To determine F_1 and F_2, we use a 'virtual source at infinity', constructed from the same back-lit reticle source as above (figure 2.7), situated in the focal plane of a converging lens L_0 ($f_0 \approx 50$ mm). The simplest way to ensure that the reticle is indeed in the focal

---

[3] A zoom lens usually has three or more components, so that the back focal plane F_2 can remain invariant while the effective focal length is changed.

plane of L_0 is to use the lens to project a real image of the source on a wall at a known large distance, and then to correct the lens position by calculating by how much to move it closer to the source in order to bring the image to infinity. For example, suppose that the focal length $f_0 = 50$ mm (20 dioptre), and the image is projected onto a wall at 2 m distance. Then, since the required change in $(\frac{1}{v})$, from $v = 2$ m to infinity, is $-1/2$ diopter and $\delta(\frac{1}{v}) - \delta(\frac{1}{u}) = 0$, the change $\delta(\frac{1}{u}) = -\frac{\delta u}{u^2} \approx -\frac{\delta u}{f_0^2} = -\frac{1}{2}$ diopter. So $\delta u = \frac{1}{2}f_0^2$ $=0.0012$ $m = 1.2$ mm, which is the lens movement required to project the image to infinity. Remember that according to the Cartesian definition u is negative, so the required change brings the reticle closer to the lens, as expected. When this device is mounted at the end of the optical rail, the image-finder can be employed to find the focal planes of the telephoto, with the combination mounted in both orientations (figure 2.14(a) and (b)). The positions of the focal planes should be marked on the cardboard strip. This will emphasize how much closer F_2 is to the lens positions than is F_1. Following this, a series of object–image data should be acquired, using the object source (figure 2.7), but now the distances to the focal points OF_1 and F_2I should be measured (using the card) rather than distances to the lenses. This is not so accurate, but much easier. Finally, the effective focal length can be found by fitting the data to Newton's law, $OF_1 \cdot F_2I = f_{\text{eff}}^2$. Following this determination, the principal points can be located and marked on the card.

2.4.4 Nodal points

The definition of a nodal point is a point around which the lens can be rotated without affecting the position of an image in the focal plane. When the system is in air ($n = 1$) the nodal and principal points coincide. We want to illustrate this by rotating the lens about one of the principal points and showing that the image of the 'source at infinity' remains static. This is a simple implementation of an experiment devised by Kingslake [1] (figure 2.15). As before, the source is at the end of the optical rail. Now the telephoto is mounted on a short extra rail which is itself

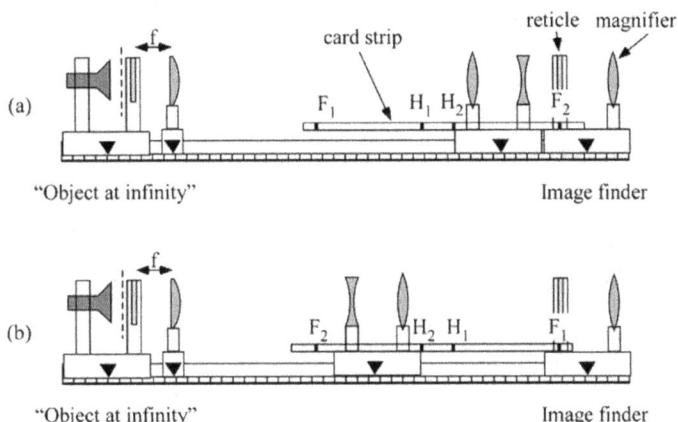

Figure 2.14. Determination of the focal planes of the telephoto combination.

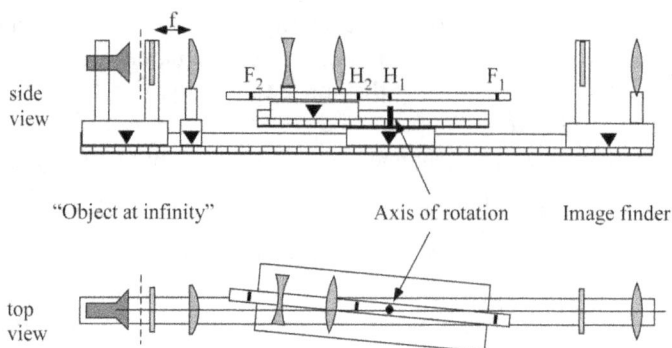

Figure 2.15. Rotating the telephoto about a nodal (= principal) point.

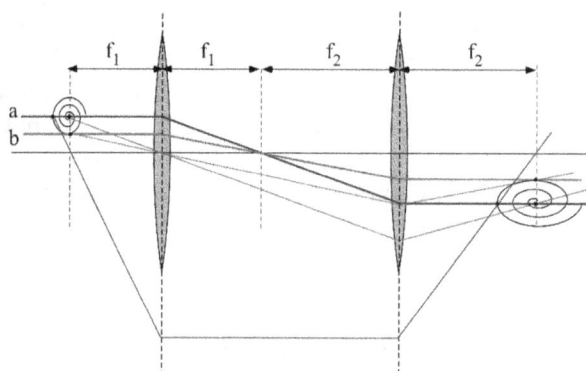

Figure 2.16. Telecentric imaging. Note that the lateral resolution is $-f_2/f_1$, but the longitudinal magnification is $(f_2/f_1)^2$.

mounted on a carrier so that it can be rotated about points in the region of H_1. We observe the image at F_1 using the image finder (the card strip may have to be folded out of the way in order to do this). When the extra rail is rotated by small angles about a point which is not exactly H_1, the image translates sideways, but if the rotation is exactly about H_1, the image remains stationary. The telephoto can be reversed and the experiment repeated for rotations about H_2, while observing at F_2.

2.4.5 Telecentric lens combination

Another interesting lens combination is the telecentric system, which has applications in metrology. This is essentially a telescope used as an imaging lens; two converging lenses are separated by the sum of their focal lengths, and an object is placed in one of exterior focal planes, to produce an inverted image in the other external focal plane (figure 2.16). This combination is actually afocal ($f_{\text{eff}} \to \infty$) and it has lateral linear magnification $-f_2/f_1$, independent of the object position along the axis, which can be seen directly by following the chief rays, labelled a and b in the figure. You should prove this property using paraxial matrix optics. That is why the telecentric system is convenient for metrology, because if the object is

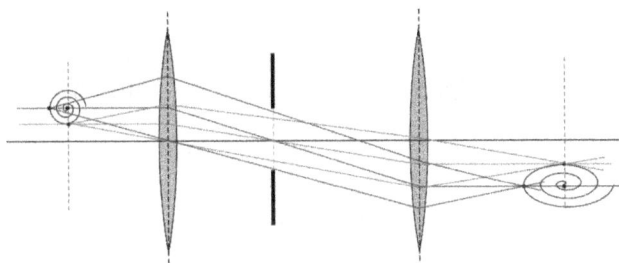

Figure 2.17. The telecentric system with a stop in the common focal plane.

three-dimensional, the lateral image dimensions are still reliable, even if they correspond to features in different planes. There is obviously a problem of depth of focus, but this can be alleviated by introducing a stop in the common focal plane; the smaller the stop, the longer the depth of focus but at the price of worsening the resolution (figure 2.17).

2.5 Telescopes: refractor telescopes, Newton reflector telescope and periscope

In this experiment we can build several types of telescope and investigate their properties:
   a. Basic refractor telescope.
   b. Refractor telescope with a field lens.
   c. Refractor telescope with an erector lens (terrestrial telescope).
   d. Galilean telescope.
   e. Newtonian reflector telescope.
   f. Periscope.

2.5.1 The concepts of stops and pupils

Initially, it is useful to remember several concepts which are common to coaxial imaging instruments, in particular telescopes and microscopes.
   a. *The aperture stop.* This is the aperture which limits the amount of light received from an on-axis point source. It is usually designed to be the aperture of the most critical or expensive optical component, e.g. the aperture of the objective (telescope or microscope) or maybe that of a scanning mirror in a scanning imager.
   b. *The entrance pupil.* This is the image of the aperture stop in all the optics (if any) coming before it. In this plane one would calculate the illuminance received from the source.
   c. *The exit pupil.* This is the image of the aperture stop in the optics following it. In a visual instrument, it would be the optimum position to locate the iris of the observer's eye.
   d. *The field stop.* This is an aperture which limits the field of view, either angular or spatial. It may have a reticle scale attached to it in a measuring

instrument. The field stop can have several purposes. For example, it may be used to limit the field of view to that for which aberrations are sufficiently small, or for which internally scattered light is not a problem.

The three stops a–c defined above are all images of one another, and if there is no absorption or scattering within the instrument, the illuminance (watt per unit area per unit solid angle) should be the same in all three.

2.5.2 Refractor telescope

The original refractor telescope (figure 2.18), patented by Hans Lippershey in 1608, is constructed from two converging lenses, with objective focal length f_1 (say 400 mm) and eyepiece f_2 (say 50 mm) which provides telescope angular magnification $m = -f_1/f_2$. In a real-life telescope, the objective is both aperture stop and entrance pupil. When the magnification is large, the eyepiece has a much smaller diameter than the objective; this is relevant to the experiments relating to the field of view, so a suitably smaller lens should be chosen, or a circular aperture or an iris should be attached to a larger lens (figure 2.19, centre box). The ratio of the eyepiece diameter to that of the objective should be about $-1/m$, or larger; otherwise, if it is smaller, the eyepiece becomes the aperture stop, which is a waste of the size of the objective. A field reticle reference scale can be introduced into the common focal

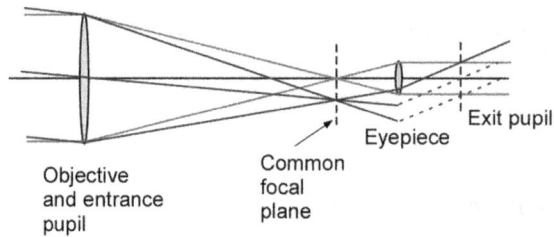

Figure 2.18. Refractor telescope. The exit pupil is the image of the objective formed by the eyepiece.

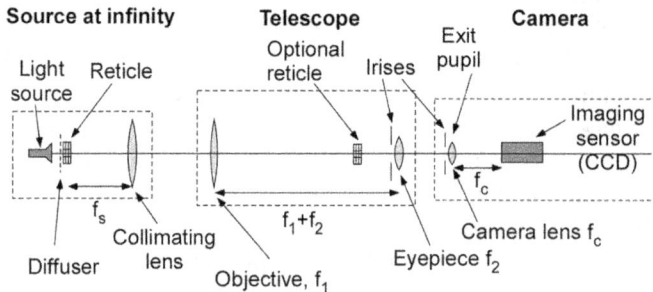

Figure 2.19. Optical layout for the telescope measurements. You may notice that the layout is the same as that of a microscope observing the source reticle (see figure 2.25), but the significances are different.

plane of the objective and eyepiece, where it will be seen sharply superimposed on the image and can be used for measurement purposes.

Angular magnification, exit pupil and depth of field. To get quantitative results for the magnification, two things are necessary:

First, the eye should be replaced by a video camera, which is basically a unit consisting of an image sensor in the focal plane of a lens of focal length $f_c = 50$ or 75 mm (figure 2.19, right box). The camera lens aperture should be stopped down to be smaller than the eyepiece diameter, to represent the eye's iris.

Second, the target projected to infinity of section 2.4.3 should be used as an object; this is a reticle scale in the focal plane of a lens (figure 2.19, left box). Calculate the focal length f_s needed for this lens to give several reticle scale divisions within the field of view in both the magnified and unmagnified images, in order to make the measurement accurate. To project the target to infinity, the lens is first adjusted to focus a sharp image of the reticle on a distant surface. Then, the additional movement needed to project the image to infinity is calculated as in section 2.4.3 and the lens moved that much closer to the reticle. To avoid over-exposure, a diffusing screen (piece of paper is fine) might be added between the light source and the reticle. If the target is not projected to infinity, the magnification measured is not accurate.

The left and right boxes in figure 2.19 should first be constructed on an optical bench 1.5–2 m long. First, the camera is focused on the target, *without the telescope*, and a reference photograph is taken. Then the refractor (or other) telescope (centre box) is constructed in the centre block and its length adjusted for a sharp image, which is also recorded. The *angular magnification* can then be deduced by comparing the two photographs. Observe whether the image is inverted.

The *exit pupil* is the position where the camera lens (or the observer's eye) should be placed in order to maximize the field of view. The camera lens should be stopped down, as explained above, in order to make this demonstration convincing (remember that the eye lens has maximum diameter about 5 mm). In a visual telescope, the exit pupil is the optimum position for the observer's eye, and is often located on a telescope by a rubber disc (called 'eye relief'), so that when you contact your eye-surround to the disc, it is automatically in the optimum position. It is instructive to take several pictures of the field of view as the camera is moved through the exit pupil.

To measure the *depth of field*, first put a mechanical stop on the optical bench which will allow the lens f_s to be moved, and afterwards to be returned to its correct position. Then f_s can be moved to simulate an object at a finite distance until the image starts to be noticeably blurred, which defines the depth of field. After a short calculation (the same as shown above) the finite distance to which the object reticle is projected can be calculated and compared with a calculation. Note that the depth of field is independent of the object distance if it is expressed in diopters. The result is a function of the telescope magnification and the objective aperture diameter. It is worth giving some thought to designing an object for which 'noticeably blurred' can be well-defined.

2.5.3 Field of view

Figure 2.20 shows the idea of the *field lens*, which in principle increases the field of view and makes the exit pupil coincide with the eyepiece. In order to make this clear it is necessary that the eyepiece diameter be considerably smaller than that of the objective (ratio about $-1/m$, as described above), otherwise the effect of the field lens is not convincing. Without the field lens, the exit pupil can be located visually in the plane where the field of view is maximized. The field of view does not then have a clean edge, and is limited by the eyepiece diameter. When the field lens is introduced, it both increases the field of view, and its edges define a sharp *field stop*. Also, the exit pupil becomes coincident with the eye lens. Of course, you may not have a lens available with exactly the focal length needed to image the objective on the eyepiece, and so the exit pupil may not exactly coincide with the eyepiece (which may not be a disadvantage) and the field stop may not be sharp.

2.5.4 Terrestrial telescope

When an *erector*, or *relay lens* is added (figure 2.21), the telescope magnification is dependent on its exact position, but basically it reimages the intermediate image with a linear magnification which is approximately -1. The erector lens often introduces some distortion and reduces the angular field of view. You should verify

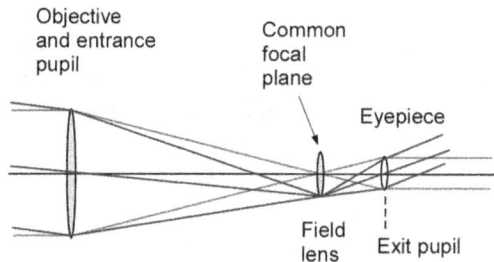

Figure 2.20. Indicating the idea of the field lens. The red and blue ray bundles in figure 2.18 showed imaging *without* the field lens. It is clear from that diagram that the blue rays were mainly vignetted by the small eyepiece (the dashed rays were not passed by the eyepiece, but were used to define the exit pupil). On the other hand, when the field lens is introduced, all the rays, red and blue, go through the eyepiece, which now coincides with the exit pupil, so the field of view is expanded. The angular size of the field of view is determined by the diameter of the field lens, which becomes the field stop.

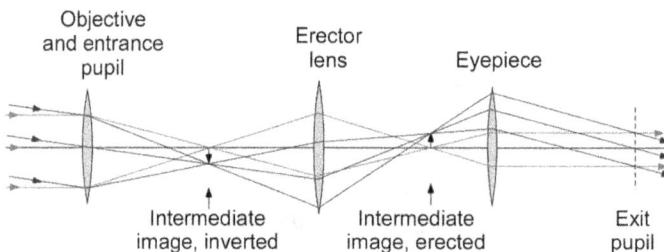

Figure 2.21. Terrestrial telescope, using an erector lens, has positive angular magnification.

that the exit pupil is now considerably further from the eyepiece than in the case of the basic refractor telescope. If a field reticle is added, there are two possible planes for it; which is preferable, and why?

2.5.5 Galilean telescope

The Galilean telescope (figure 2.22) uses an eyepiece with a negative focal length, and produces an upright image ($m = -f_1/f_2$ is positive). Since the telescope length is $f_1 + f_2$, it is shorter than the equivalent refractor telescope built with the same objective and a converging objective having the same magnification. However, the exit pupil, which is the image of the objective in the eyepiece, is now a virtual image; the observer's eye-pupil cannot coincide with it and so the field of view is severely limited. Also, a field reticle cannot be incorporated, because there is no real intermediate image plane. A popular use of the Galilean telescope is for opera-glasses. In this case, the positive magnification and short length are important, and the smaller field of view can not only be tolerated, but is often an advantage!

2.5.6 Newtonian reflector telescope

Reflecting telescopes have the great advantage of being achromatic, and so it is worth using an achromatic doublet for the eyepiece. The Newtonian (figure 2.23) is the only reflecting telescope which can be constructed using standard optical elements, since it does not require an objective mirror with a hole in it, as do the Cassegrain, Gregorian, Ritchey–Chrétien and other designs. But, just the same, the diagonal mirror of the Newtonian does obscure some of the aperture stop. Note that the diagonal mirror is elliptical and is mounted slightly off axis to get minimum obscuration for a given angular field of view; special elliptical plane mirrors for this purpose are standard off-the-shelf components.

2.5.7 Periscope

The periscope is the 'ultimate' illustration of the idea of a field lens. Basically, a periscope consists of two mirrors at angles $45°$ to the optical axis (figure 2.24(a)), separated by distance L. The construction includes a tube of radius R which prevents external light (or water!) entering the system. This system has a paraxial angular field

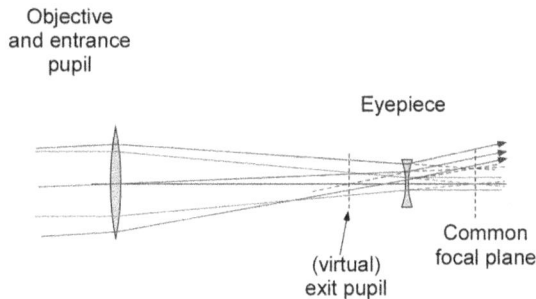

Figure 2.22. Galilean telescope. The exit pupil is not accessible to a camera or eye.

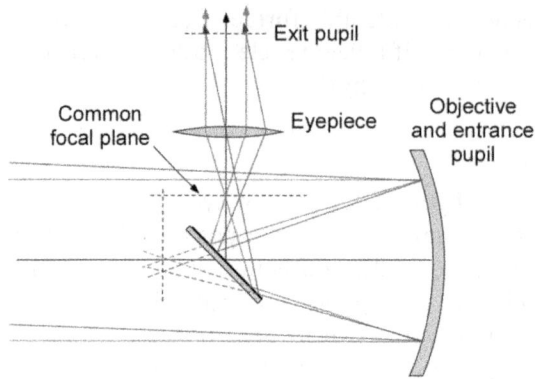

Figure 2.23. Newtonian reflector telescope.

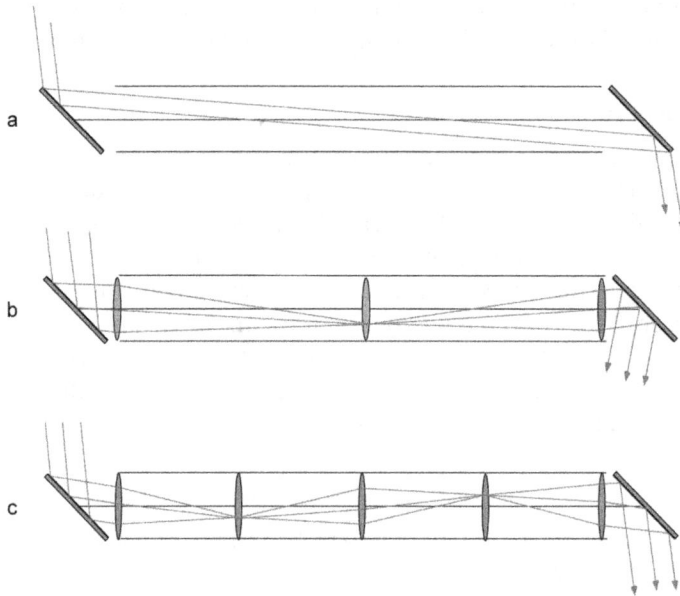

Figure 2.24. Schematic diagrams showing (a) a periscope with no lenses; (b) the effect of adding a unit magnification telescope with field lens (inverts the image); (c) adding two more lenses to increase the field of view by a factor of two.

of view $4R/L$, although the margins of this field pass very few rays and are therefore very weak and indistinct. A first improvement would be to build a telescope with unit magnification to guide more rays through the tube, and to increase the field of view by including a field lens (figure 2.24(b)). The objective and eyepiece would both have focal lengths $L/2$, and the field lens (which images the objective on the eyepiece) has focal length $L/4$. This results in a field of view which is still $4R/L$ but has uniform intensity and sharp edges. This principle can then be continued, using the idea of the terrestrial telescope, by adding several relay lenses and more field lenses. An example of the construction is shown in figure 2.24(c) and consists of objective and eyepiece with focal lengths f_0 and alternating relay lenses and field lenses, all with focal length $f_0/2$, separated by f_0. The angular field of view is then $2R/f_0 = 8R/L$ and the number of lenses is $1 + L/f_0$. The number of lenses can be increased; in a submarine periscope it may be of order 20. Of course the image might be distorted after so many relays, if the lenses are not specifically designed for the purpose. Moreover, if the lens surfaces are not anti-reflection coated, the contrast of the image is compromised; for example, if there are 20 uncoated lenses, each surface reflecting 4% of the incident light, about 80% of incident light is reflected or scattered. If the reflection coefficient can be reduced by a broad-band coating to 0.2%, only 8% is lost. These are important considerations in designing real-life periscopes. However, in the laboratory a periscope 0.8 m long built with nine standard bi-convex lenses, seven with $f = 50$ mm and two with $f = 100$ mm gave a surprisingly clear and undistorted image with field of view eight times larger than the basic form in figure 2.24(a) and (b).

2.5.8 Compound eyepiece

The combination of field lens and eye lens, which provides optimum field of view, is a common item to many telescopes and microscopes. The field lens may be displaced from the focal plane of the eye lens, both to allow inclusion of a measuring reticle in that plane and also to avoid dust on the lens affecting the image. You will often find them combined as a single unit; such a combination is called a 'compound eyepiece', and is often labelled with its magnification, based on its use in a microscope with a standard length (see section 2.6). An advantage of such a compound eyepiece is that, having two lenses or more, it can be designed to eliminate chromatic aberration.

2.6 Microscopes: transmission, reflection, dark field

The first microscopes similar to those in use today were built by Robert Hooke in England and Anton van Leeuwenhoek in Holland in the mid-seventeenth century. Since that time, they have been continuously improved, regarding both their spatial resolution and the variety of object properties which can be observed. Since microscopes can resolve detail in details of an image smaller than the wavelength of the light used for the observation, it is obvious that physical optics are needed to understand them in detail; this was first realized by Ernst Abbe who, in 1873, related microscope resolution to the wavelength of light with the famous formula $d_{min} = \lambda/2\text{NA}$, where NA is the numerical aperture of the imaging lens (section 2.6.3). Since then there has been a continuous effort to provide 'superresolution' in excess of this limit, for which three researchers received a Nobel Prize in 2014, as well as the employment of radiations such as x-rays and electrons which have wavelengths much smaller than those of visible light. Another recent advance, called 'Ptychography' [2] and presented in section 4.7, achieves high resolution within a large field of view, by combining several images taken with different angles of illumination through a low magnification objective. But the basic element in any microscope is a very high quality objective lens. The design of such lenses is a major industry, but it is significant to note that Hooke's microscope of 1665 used a drop of honey as the objective. It is an interesting laboratory project to try to reproduce this achievement; it fits very well with a major modern discipline called 'free-form optics'!

2.6.1 Construction

The experiments we discuss here will lead to a basic understanding of the principle of the optical microscope in several different forms. We construct a model microscope using two converging lenses: a short focal length (30–50 mm) as the objective and a longer focal length (150–250 mm) as the tube lens. The two lenses are separated by a distance of about 60 mm. The object is in the back focal plane of the objective, and the image is recorded on a camera sensor placed in the focal plane of the tube lens. The objective and tube lenses should be *achromatic doublets*; this makes a lot of difference, even with quasi-monochromatic light, since achromats have some correction for spherical as well as chromatic aberration. In addition, the orientation of the lenses is important, i.e. which surface faces the object and camera since achromats do not have a symmetrical geometry (see figure 2.25). An iris diaphragm in the back focal plane of the objective (Fourier plane) serves to control its numerical aperture, which affects the resolution, but the physical optical aspects of imaging and resolution are better illustrated in the experiment on coherent imaging (section 4.4.1). Although the construction principles are best understood using the above achromat as objective, it is worth changing this to a professional objective at a later stage in order to see some of the problems introduced on going to higher magnification, such as aberrations and field of view limitations.

The illuminator for the microscope can be:
1. a collimated laser beam (see the experiment on coherent imaging, section 4.4),
2. a flashlight followed by a ground-glass plate, the simplest device for incoherent imaging (figure 2.25).
3. a Köhler illuminator, consisting of a light source focused onto an iris diaphragm in the back focal plane of the condenser lens. This provides uniform incoherent illumination with a degree of spatial coherence that can be controlled by the diameter of the iris (figure 2.26).

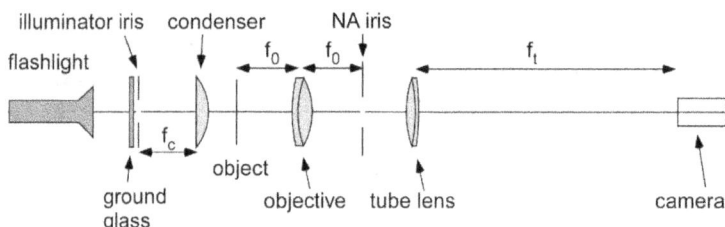

Figure 2.25. Microscope layout with a basic illuminator.

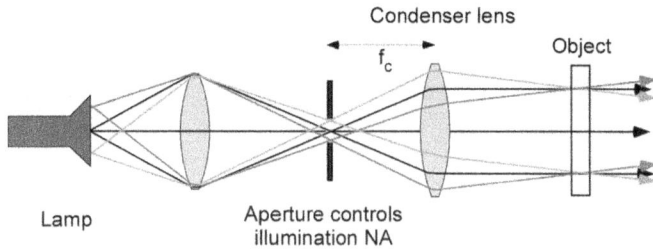

Figure 2.26. A Köhler illuminator. Note that the object plane is conjugate to the lens on the left, which is uniformly illuminated by the lamp.

2.6.2 Magnification

The magnification can be measured using a reticle as object, and calibrating the camera by putting the same reticle close to the camera plane. It is equal to f_t / f_o when the object is in the focal plane of the objective. In the special case when the distance between the objective and the tube lens is equal to the sum $f_o + f_t$, (telecentric microscope, section 2.4.5) this magnification is maintained even as the object goes out of focus, but the large separation of the lenses makes the microscope inconveniently long, and also the field of view becomes very small. The telecentric system is sometimes employed in metrological microscopes with small magnification.

2.6.3 Numerical aperture

The numerical aperture (NA) of a microscope objective is defined as the sine of the half-angle of the cone defined by the aperture stop and the centre of the focal plane, multiplied by the refractive index of the medium in that region (which might be air or an immersion fluid or gel). The Abbe limit of resolution is $\lambda/2NA$. In general, higher magnification objectives have larger NA, and can therefore give better resolution. One of the achievements of ptychography (section 4.7) is to get high resolution using a low magnification objective. In the experiment of figure 2.25, the NA can be controlled by an iris in the focal plane of the objective; although this is not in the aperture stop, its effect is the same and can be more easily dealt with analytically.

2.6.4 Depth of focus

When we observe an object with three-dimensional structure, it is clear that not all sections of the image can be recorded sharply by a two-dimensional sensor. The depth of focus is the allowed variation of the object plane along the axis for which the image appears sharp. Using an object with some fine detail, which can be moved along the axis, you can get an impression of how the depth of field varies with the with the NA; from geometrical optics, one would deduce that the depth of field is proportional to $1/NA$, but from physical optics it is proportional to $1/NA^2$, since the spatial resolution also depends on the NA.

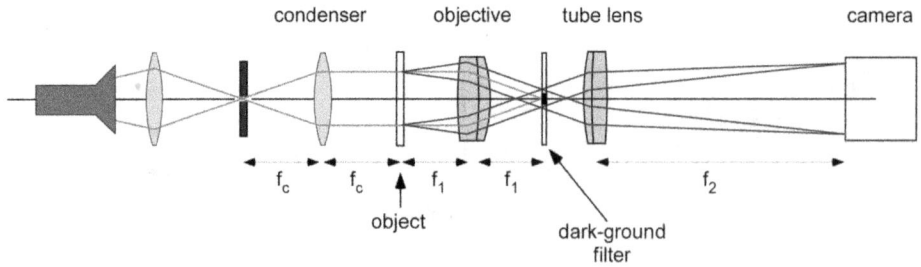

Figure 2.27. Dark ground microscope using a Köhler illuminator. All rays (red) which are not scattered by the object are stopped by the dark-ground filter, but scattered rays (blue) bypass the filter.

Figure 2.28. Images of a reticle with marks at 0.1 mm intervals using the setup of figure 2.27: (a) bright field image, without the dark-ground filter, and (b) the dark-ground image; notice that the image is dominated by edges of the lines and dust on the reticle, which scatter the incident light.

2.6.5 Dark-ground imaging

The idea of dark-ground imaging is to emphasize small opaque details on a mainly bright background. To illustrate the idea, we can replace the NA iris by a glass slide with a circular obstruction (e.g. a drop of black ink which has been allowed to dry) on the axis. Using Köhler illumination (figure 2.26), without an object, the illuminator iris is stopped down till its image in the NA plane exactly coincides with the black spot, so that very little light reaches the camera directly. Now introduce a mainly transparent object such as a reticle; the image will now emphasize the parts which scatter light, such as small details or edges (figures 2.27 and 2.28). Dark-ground imaging using coherent illumination is discussed in section 4.4.4. The system can be made to have higher resolution and to be more light-efficient by using an annular aperture in the illuminator and a matching ring stop, which are implemented in commercial instruments, but these require very careful construction to be effective.

2.6.6 Reflection microscope

When the object is opaque, the construction of the microscope must be modified so that light enters through a beam-splitter and the objective lens, and is reflected back through the objective and tube lens to the camera (figure 2.29). The beam-splitter, which may be simply a thin plate of glass at 45° to the axis, has to be situated

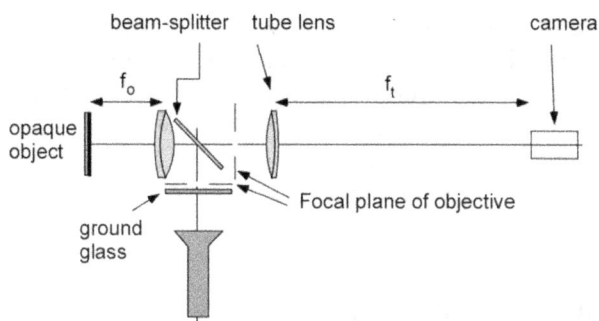

Figure 2.29. Reflection microscope.

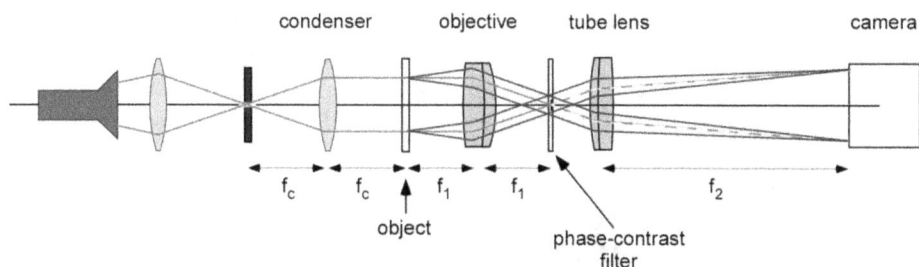

Figure 2.30. Zernike phase-contrast microscope. The dashed red rays are phase-shifted by 90° and interfere with other rays (blue) from the same point on the object.

between the objective and its back focal plane in order to allow dark ground and other spatial filtering techniques to be implemented.

Metallurgical specimens are commonly observed with reflection microscopy, and in this experiment a piece of galvanized iron or other polycrystalline surface makes an interesting object. The dark ground technique can be used here too.

2.6.7 Polarization and phase microscopy

One type of functional microscopy emphasizes spatially-dependent polarizing properties of the object, which could be collection of small sugar crystals, for example. In this case the illuminator contains a linear polarizer, and there is an orthogonal polarizer after the objective.

Another type of functional microscopy emphasizes phase variations from point to point in the image of an otherwise transparent object. These might be the result of refractive index or thickness variations or even applied fields. The first such microscope was built by Zernike in 1934 and essentially revolutionized biology by allowing detailed microscopy of living cells without the use of staining to distinguish between the different materials in their structure; staining often altered their behaviour or even killed them. Zernike was awarded a Nobel Prize for his invention in 1953. A challenging experiment is to construct a Zernike phase-contrast micro-scope based on the same system as the dark-ground microscope (figure 2.30). In this case, the principle can be implemented if the circular obstruction on the glass slide is replaced by a disc of transparent material with thickness d and refractive index n such that the optical thickness difference from the surroundings is $d(n - 1) = \lambda/4$ at the centre of the visible spectrum. One might use a drop of diluted transparent lacquer which has an appropriate thickness when it dries; the thickness does not need to be very accurate to demonstrate the idea. Then of course a suitable object has to be provided, which is transparent but has varying optical thickness from point to point.

2.7 Autocollimator: measuring focal planes of a lens and angle of rotation

The auto-collimator is a useful optical instrument which enables measurement of focal lengths, accurate measurement of angles and alignment of optical set-ups. The basic layout is shown in figure 2.31(a), and consists of an illuminated source reticle 1, followed by a beam-splitter and a converging lens L_1. At the second exit from the beam-splitter are situated reticle 2, in the plane coincident with the image of reticle 1 in the beam-splitter, and a magnifier to observe it. When a plane mirror M is placed normal to the axis, a sharp inverted image of reticle 1 is formed on reticle 2 when the lens L_1 is correctly positioned so that reticle 1 is in its front focal plane, implying that the light between L_1 and M is collimated; the distance between L_1 and M is immaterial to the sharpness of the image, but affects the field of view. The reticle structure is not critical; it is convenient to use for reticle 1 a fairly coarse high contrast sharp-edged grating (say 0.2 mm period) so that its image can easily be seen,

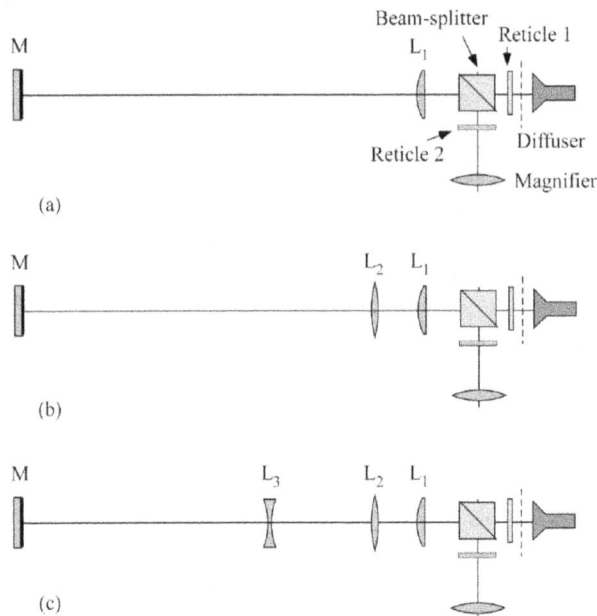

Figure 2.31. The autocollimator. (a) Basic construction; (b) used to find the focal plane of a converging lens; (c) used to find the focal plane of a diverging lens.

blurred, even if it is far from focus. If L_1 is not achromatic, it will be necessary to add a colour filter to get a really sharp image. An image can also be observed if the system is misadjusted so that L_1 creates an image of reticle 1 on M; this case can be recognized by the fact that the distance to the mirror is critical and the image is upright, rather than inverted, and it should be avoided.

In the configuration of figure 2.31(a), the autocollimator can be used to accurately measure angles of tilt of the mirror, by observing lateral shifts of the image of reticle 1 with respect to reticle 2. Estimate the accuracy that can be claimed for such a system.

The autocollimator is also employed to find focal planes and focal lengths of thin lenses conveniently. If the converging lens L_2 is inserted (figure 2.31(b)), an upright image is obtained when the mirror is in the back focal plane of L_2. So the focal length of the lens can be found, assuming it is a thin lens. Following this, the focal length of a stronger diverging lens L_3 can be found by finding where to place it relative to the mirror so that the original inverted image is restored (figure 2.31(c)). Then, $-ML_3$ is the focal length of L_3.

It is convenient to have an autocollimator permanently set up in an optics lab, so that focal lengths of converging and diverging lenses can quickly be checked. Of course, one has to remember that the instrument locates the focal planes, not the focal length, and to determine the focal length the principal planes have to be known too. But for singlet lenses, the assumption that the principal planes are close to the central plane of the lens is good enough for most purposes.

Another interesting experiment is to estimate the focal length of a 'meniscus lens' (one surface convex, the other concave with larger radius). The lens may be thin, but the principal planes, although they more-or-less coincide, are outside the lens. On which side, the concave or convex one?

2.8 Aberrations and their reduction: some basic concepts, use of stops

The study of lens aberrations is rather technical and usually relates to very specific optical designs, but there are some aspects which are quite general and are worthwhile understanding. These are:

1. Chromatic aberration.
2. Spherical aberration: Its elimination by use of the Abbe sine rule (the aplanatic lens), or by commonly-used aspheric lenses; the choice between two orientations of an asymmetric lens (plano-convex, plano-concave, achromatic doublets).
3. Astigmatism and coma at field points distant from the axis.
4. Distortion in optical systems used for linear measurement.

2.8.1 Chromatic aberration

Chromatic aberration of a lens results from the dependence of the refractive index of transparent materials on the wavelength. This aberration can be treated using paraxial optics, and the outcome is the design of astigmatic doublets made by cementing together two lenses of opposite focal lengths made from glasses with different dispersions (section 2.1) [3]. The compound lens is designed so that its effective focal lengths at a red wavelength and a blue wavelength are equal. But the focal length in the green is slightly different; can you design an experiment which measures this difference? Is an auto-collimator (section 2.7) sufficiently sensitive for this?

2.8.2 Spherical aberration

Spherical aberration is the predominant aberration in lens systems which use rays which are not paraxial. If we try to make the angles of incidence and refraction at the various surfaces as paraxial as possible, the aberration will be minimized. As a result, a general principle for reducing spherical aberration is to try to ensure that the deviations of a given ray are, as far as possible approximately equal at the successive surfaces. This can be well illustrated using a plano-convex lens to produce an image of a distant object on its axis. Then, if the plane surface faces the object, the rays from it are incident at zero angle, and all the deviation occurs at the second surface. On the other hand, if the curved surface faces the object, the deviation is divided between the two surfaces and the spherical aberration is less. This can be confirmed experimentally.

When achromatic doublets are designed, three radii of curvature are involved, but there are only two equations. So there is a free parameter which is used to choose the radii of curvature of the surfaces to minimize spherical aberration. The result is a ratio of approximately[4] 5 between the curvatures of the outer surfaces; again, the more curved surface should face the distant object. For this reason, it is a good idea to use an achromatic doublet for sharp focusing of a collimated laser beam, even though the light is monochromatic!

2.8.3 Off-axis aberrations

When an off-axis point in the field of view of an imaging system is imaged, coma and astigmatic aberrations can be introduced which depend on the cube of the distance from the axis. These are obviously important in wide-field imaging systems. Coma, as its name suggests, makes the image of a point source look like a comet with a bright head and a weaker tail in the radial direction, and is an aberration which is important in small f-number (large aperture) systems. Astigmatism, which results in a point image becoming a line (a caustic, in wave theory), occurs with even small apertures, and is a signature of a lens misaligned with respect to the optical axis of an arrangement[5]. We can see this in a very simple experiment in which a collimated broadened laser beam is incident on a lens at an angle to its axis. The light focuses to a line on a screen close to the focus in two different planes; one line is in the plane of incidence, and one is normal to it. And the distance between the two planes is of course a function of the angle between the beam and the axis. For example, we used a lens with diameter 25 mm and focal length 175 mm with its axis about 30° to a

---

[4] For a singlet lens, it is easy to show that for equal deviation, the ratio between the curvatures is $n/(2 - n)$ by considering the lens approximately as two prisms cemented base to base.

[5] Astigmatism, or cylinder, is also well-known in ophthalmic optics, where it can also occur for an axial point. But there it is related to a distortion of the eye lens, which has different radii of curvature in different planes of incidence.

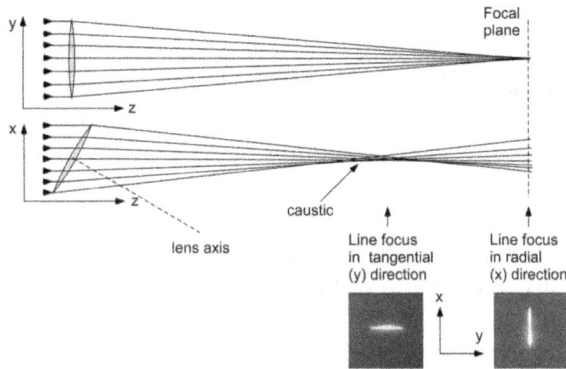

Figure 2.32. Two line images created by astigmatism.

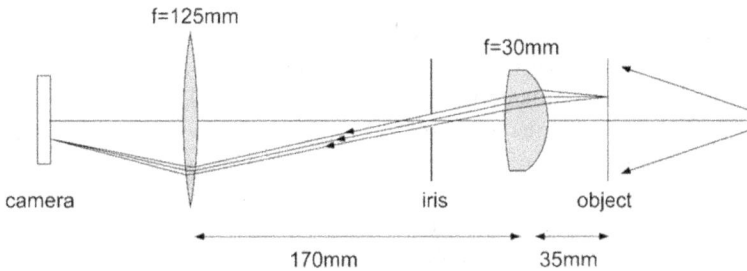

Figure 2.33. Experiment to illustrate distortion in an imaging system.

collimated beam (see figure 2.32) and found focal lines in two planes separated by about 30 mm, which is not small compared with the focal length! As the angle of incidence decreases, the planes of the two line images converge on the paraxial focal plane, and the lines get smaller in length. These observations can be measured systematically as a function of the angle and can be compared to non-paraxial calculations for a single lens.

2.8.4 Distortion

Even if a lens system creates a perfect image in the sense that a point object is imaged to a point in the image plane, the magnification may be a function of the point's position in the field of view. If this is so, the shape of an object may be distorted in the image. For example, if the object is a square, then if the magnification depends on the position, the length of the diagonal of the square will not be exactly equal to $\sqrt{2}$ times the side. If the magnification grows with distance from the axis, the diagonal is longer than that of the square and the distortion of the resulting image is called 'pincushion distortion', while if the magnification reduces with distance, we have 'barrel distortion'. These situations can be demonstrated by using a lens to image a suitable slide with magnification not equal to unity and selecting the group of rays used for imaging by means of an axial aperture at a distance on one side of the lens or the other [3]. Note that if the magnification is unity, a simple symmetry argument shows that there can be no distortion. Distortion is very prominent when imaging with an asymmetrical system such as a telephoto combination.

An experiment to show the influence of the position of the axial aperture on the degree of distortion is shown in figure 2.33. A piece of semi-transparent mm graph-paper, which acts as both object and diffusing screen, was illuminated by a broad expanding laser beam (this is for convenience; the scattering makes the object illumination incoherent). It was imaged by an achromatic doublet with focal length 30 mm situated at a distance a little more than its focal length so that a real and magnified image was created at a distance of about 200 mm—a very asymmetrical arrangement. The orientation of the lens, which is also asymmetrical, was as shown. The image was relayed to a camera by means of a weak lens which reduced its size on the sensor. A coaxial iris aperture about 2 mm diameter was moved along the axis in the image space. The images (figure 2.34) show that when the iris was close to the lens (a) the image was essentially undistorted, but as it moved away pincushion distortion was observed which increased with the distance from the lens (b,c,d). In this experiment it is important that the lens aperture (25 mm) is large enough to accept rays at angles well outside the paraxial region. The degree of distortion (linear magnification as a function of angle of incidence) should be compared to a ray-tracing model. Because of the short distance between the object and the lens,

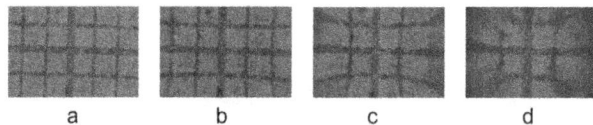

Figure 2.34. Increasing pincushion distortion with the distance between the axial aperture and the lens on the image side: (a) 20 mm, (b) 50 mm, (c) 70 mm, (d) 90 mm.

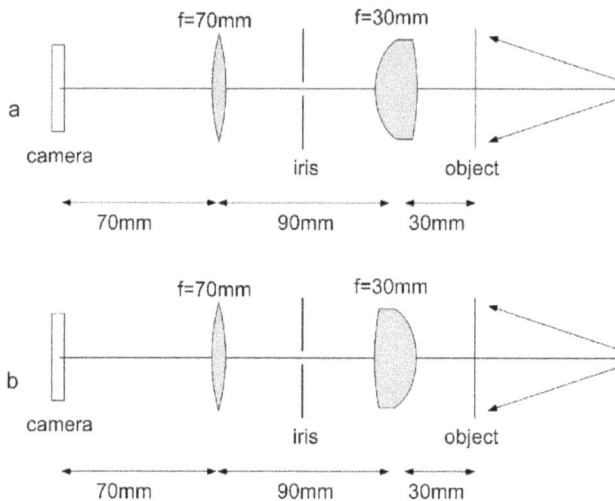

Figure 2.35. Influence of the orientation of a biconvex lens on distortion in a magnifying system: (a) the more-preferred orientation, (b) the less-preferred one.

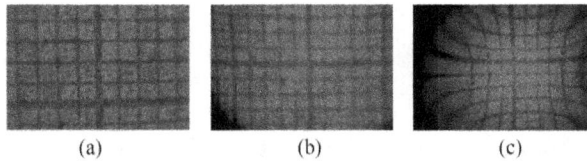

Figure 2.36. (a) and (b): images corresponding to figure 2.35(a) and (b), respectively. (c) corresponds to figure 2.35(b) with the aperture between the 70 mm lens and the camera.

it was difficult to repeat the experiment with the iris placed on the opposite side of the lens. In that case barrel distortion should be observed.

The influence of the *orientation of the lens* on distortion is another subject which can be investigated. In order to make a magnified image for measurement purposes, one can use a single lens (either biconvex or plano-convex, depending on the magnification required) or a pair of lenses with focal lengths in the required ratio. The latter is more accurate because the curvatures of lens surfaces are usually arranged so as to provide the best image when one conjugate is infinite. We took two lenses with focal lengths 30 mm and 70 mm separated by about 90 mm with a 4 mm axial aperture half way between them to get a magnification of 7/3, and compared the distortions when the more curved surfaces were facing (figure 2.36(a)) and when the more planar surfaces were facing (figure 2.36(b)). When the aperture was placed outside the lens pair, near the camera, the distortion was considerably greater (figure 2.36(c)).

2.9 Gravitational lens analogy: an example of an aspherical lens

Lenses with spherical surfaces are ubiquitous in geometrical optics, but are only a paraxial approximation to a lens with ideal properties. They show aberrations of many types under non-paraxial conditions. Most advanced lens designs use combinations of spherical-surfaced lenses to counter the aberrations, since these lenses are relatively easy to manufacture with high quality. Aspherical lenses, with surfaces that are not spherical, have many applications in optical design and today are quite widely used in mass-production situations, when the higher cost of their manufacture is compensated by their simplicity (e.g. cell-phone cameras). Because they are designed for specific purposes, it is difficult to suggest an interesting lab experiment which illustrates typical properties of such lenses. Nevertheless, Nature provides us with a very interesting type of aspherical lens, which has had important consequences in observational astronomy and is basically quite easy to understand. This is the gravitational lens, created by the gravitational field around a massive stellar object.

2.9.1 Gravitational lensing

When a light ray passes near a massive body, the light ray bends. The reason for the bending is that, according to general relativity, the structure of space-time becomes distorted by the massive body [4, 5]. Since the light travels in a straight line in empty space, when it is observed in the space containing the mass, the ray becomes distorted. One way of looking at it is as if we have a freely falling lift, in which a laser beam propagates from one side to the other. As there is no effect of the Earth's gravity inside the lift (as its occupants, who are floating freely in it will tell you) the laser beam travels in a straight line from one side to the other. On the other hand, observers in the laboratory who watch the lift pass by with acceleration g relative to their frame of reference will tell you that the laser beam travels along a path which is not quite straight, since the lift was falling more slowly when the beam left the laser than when it reached the far side of the lift. In cosmology, when light from a distant star passes close to a massive body on its way to an observer on Earth, its direction becomes slightly changed, so that the apparent position of the star changes slightly.

Historically, the effect was predicted by Soldner in 1801, based on Newton's theory of light which considered a light ray as being composed of particles with mass m moving at a constant velocity c. The result was independent of the mass m, so presumably should also apply in the limit m going to zero. When a ray from the distant star passes a distance b from a spherical body of mass M on its way to the Earth, Soldner showed that there would be an angular deviation $\alpha(b) = 2MG/bc^2$. In 1916, Einstein showed that the result according to general relativity should be $\alpha(b) = 4MG/bc^2$, twice the classical value. Einstein was proved correct by Dyson and Eddington, who measured in 1919 how the apparent positions of stars changed as the Sun passed near their light paths during the total solar eclipse that year[6]; the total eclipse was needed to make the stars visible when their angular positions were close enough to the Sun to make the effect measurable [4]. Given the diameter, distance and mass of the Sun, you should estimate the angular displacement of a star when it is observed close to the Sun's rim; it is very close to the limits of optical resolution by a telescope observing through the Earth's turbulent atmosphere (about 1 arc-sec). In the 1960s, the most conclusive measurements were made using radio astronomy, which does not require an eclipse since the Sun does not have strong radio emission.

The bending effect, called 'gravitational lensing', distorts images of very distant stars when another star, the 'lensing mass', lies on or close to the line of sight to the observer (figure 2.37). The image is broken into parts or arcs, which are recognized as being related since they have the same spectrum, temporal fluctuations and red-shift, so must have originated from the same source. In many cases, the gravitational image of the composite image is brighter than the undistorted image would have been. It is an interesting experiment to design an optical lens which has the same imaging properties as a gravitational lens, and use it to reproduce some of these stellar composite images. The discussion here presents both a demonstration experiment, using an easily-obtainable lens with a qualitatively correct shape, and a laboratory experiment in which quantitative results can be obtained.

---

[6] A previous plan for an experiment was made to be carried out in 1914, when there would be an eclipse which would be total in Russia, but the First World War got in the way. Dyson and Eddington's results were somewhat controversial since they were spoiled by clouds, but were repeated conclusively in 1922.

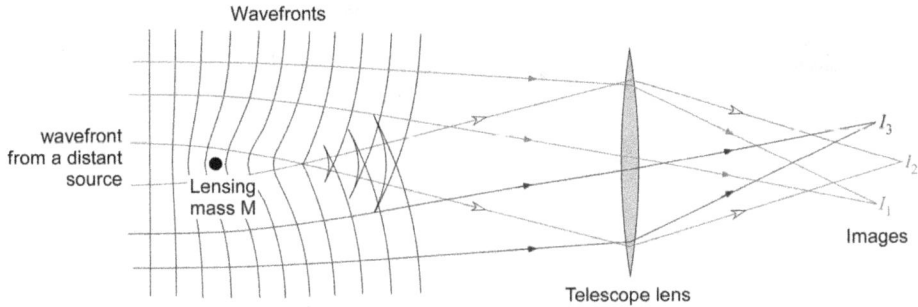

Figure 2.37. Distortion of an incident plane wavefront by the gravitational field of a massive particle results in several images. The gravitational field creates an 'effective refractive index' $n(r)$, for which $(n - 1)$ is proportional to the gravitational potential, GM/r. Reproduced from [8] with permission of Cambridge University Press © 2011.

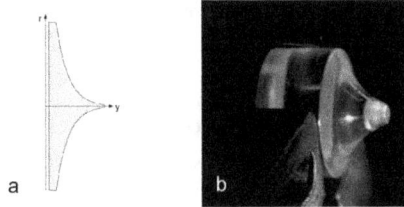

Figure 2.38. Analogue lens (a) profile, (b) a completed lens.

2.9.2 Properties of an analogue gravitational lens

Consider a plano-convex lens with thickness profile given by $y(r)$, where r is the radius from its axis. We want it to deviate a ray at radius r by an angle $\alpha(r) = A/r$. In the paraxial approximation (i.e. not for r close to zero!) we consider the lens locally as a prism with angle dy/dr. The angle of deviation is then $(n - 1)dy/dr$. Solving the equation $A/r = (n - 1)dy/dr$ we find that y must be proportional to $\log(r)$. This lens has a shape which is nothing like the lenses which we are used to, and does not focus an incident parallel beam to a focus. Figure 2.37 shows how the wavefronts of an incident plane wave are distorted by the lens. The rays can be found by following the normals to the wavefronts. Figure 2.38(a) shows the profile of a practical lens.

If the image is then recorded using a conventional telescope lens, and the lensing mass is on its axis, it is clear that the image is a series of concentric rings, which are focused in different planes. In practice, the telescope aperture (tiny, on a cosmic scale) limits us to imaging rays which pass through its centre (figure 2.39). This forms a ring image, called the 'Einstein ring', which is the signature of gravitational images. If the telescope aperture, the lensing mass and the source are not exactly on the same axis, parts of the ring are missing, and more complicated images are formed.

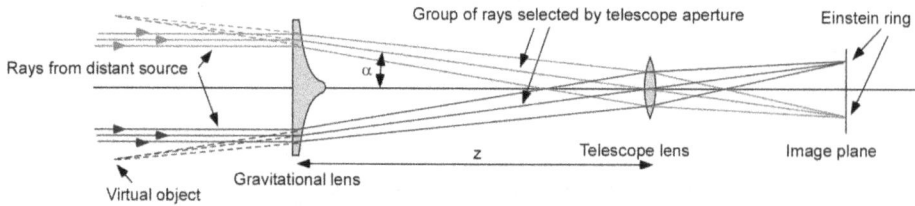

Figure 2.39. Ray diagram for the lab experiment.

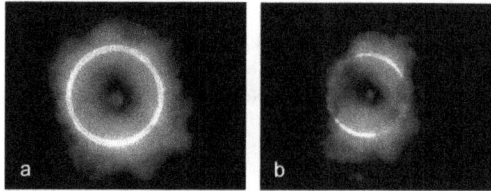

Figure 2.40. Einstein ring (a) for a source on axis, (b) slightly off axis.

2.9.3 A laboratory gravitational lens

The analogue lens with the logarithmic profile was prepared by carefully turning a Perspex rod on a lathe. First, a suitable function $y(r) = B/r$ was calculated, and then the profile reproduced by hand on a lathe in steps of about 0.2 mm in radius on a rod 12 mm diameter. Of course, one has to round off the logarithmic singularity on the axis, but the resulting bump has very small area and therefore affects very little of the light. The as-turned surface with discrete steps was smoothed and polished by hand, as well as the opposing plane surface $y = \text{const.} < 0$. The completed lens made this way is shown in figure 2.38(b)[7].

To demonstrate the Einstein ring qualitatively, you don't need to go to the trouble of turning this lens. I found that the bottom of a chemical vial bottle about 15 mm diameter has a profile which is a good enough approximation to the logarithm in its outer parts and is remarkably axially symmetric. This gave good results too; see figure 2.42.

The grav-lens images should be observed with an optical arrangement which reproduces the astronomical situation in principle. Now we assume that the rays originating from a distant source are bent only in the region of the grav-lens, so the appropriate ray diagram, figure 2.39 shows that the ring will be sharpest when the telescope is focused on a plane behind the grav-lens, which acts locally as a weak diverging lens. In astronomy, this plane is essentially at infinity, but in the lab it is at a finite distance, so the image plane is at a distance greater than the focal length of the telescope lens. The telescope lens should have a small aperture ($\sim$5 mm) for the paraxial approximation to be applicable, resulting in a sharp Einstein ring.

---

[7] In retrospect, the lens shown in figure 2.38(b) was rather thick compared with its radius, and it would have been better to make the constant B smaller.

Since the profile of the lens is known quantitatively, you can calculate the angle of the Einstein ring (figure 2.40(a)). When the size of your telescope lens is much smaller than the grav-lens (as, obviously, in astronomy) the angular radius of the ring is $\sqrt{(B/z)}$, where z is the distance between the two.

When the axis of the incident light is not exactly parallel to that of the lens, some parts of the ring are missing (figure 2.40(b)). Of course, if the lensing mass produces a spherically symmetric gravitational field, this off-axis situation must correspond to the case where the source, lensing mass and the Earth are not exactly in a straight line. As the axis deviation increases, more complicated ring-based patterns are formed; two are shown in figure 2.41, together with observational instances which are quite similar.

By way of comparison, figure 2.42 shows Einstein rings obtained from the vial base.

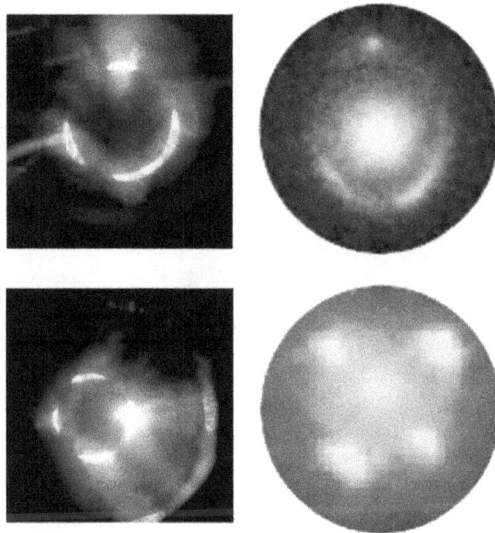

Figure 2.41. Analogue lens images compared with astronomical observations. Upper row: B1938+666 NIR image [6] observed with the Hubble Space Telescope; lower row: Huchra's Cross, Q2237+0305, in the NIR [5, 7]. Reproduced from [8] with permission of Cambridge University Press © 2011.

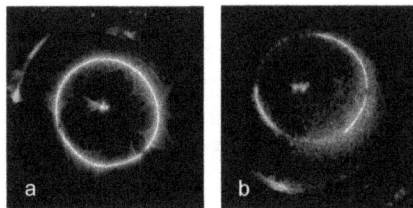

Figure 2.42. Using a glass vial base as an approximate grav-lens. Einstein rings (a) for an on-axis source, (b) off-axis.

References

[1] Kingslake R 1932 A new bench for testing photographic lenses *J. Opt. Soc. Am.* **22** 207

[2] Zheng G, Horstmeyer R and Yang C 2013 Wide-field, high-resolution Fourier ptychographic microscopy *Nat. Photonics* **7** 739

[3] Hecht E 2016 *Optics* 5th edn (New York: Pearson)

[4] Kennefick D J 2019 *No Shadow of a Doubt: The 1919 Eclipse That Confirmed Einstein's Theory of Relativity* (Princeton, NJ: Princeton University Press)

[5] Paczyński B and Wambsganss J 1993 Gravitational microlensing *Phys. World* **6** 26

[6] King L J, Jackson N, Blandford R D, Bremer M N, Browne I W A, De Bruyn A G, Fassnacht C, Koopmans L, Marlow D and Wilkinson P N 1998 A complete infrared Einstein ring in the gravitational lensing system B1938+666 *Mon. Not. R. Astron. Soc.* **295** L41–4

[7] Huchra J, Gorenstein M, Kent S, Shapiro I, Smith G, Horine E and Perley R 1985 2237 +0305: A new and unusual gravitational lens *Astron. J.* **90** 691–6

[8] Lipson A, Lipson S G and Lipson H 2011 *Optical Physics* 4th ed (Cambridge: Cambridge University Press)

IOP Publishing

Optics Experiments and Demonstrations for Student Laboratories

Stephen G Lipson

Chapter 3

Polarization and scattering

3.1 Polarized light

Since electromagnetic waves are transverse, there will always exist two independent waves traveling in a given direction, and having orthogonal fields, which are called 'linearly polarized'. One should note, from Maxwell's equations $\nabla \cdot \vec{D} = 0$ in an uncharged region and $\nabla \cdot \vec{B} = 0$, that the vectors $\vec{D}$ and $\vec{B}$ are transverse to the propagation vector $\vec{k}$. These fields are the electric displacement field and the magnetic induction field, respectively, and they define the polarization of the wave. It is significant that the polarization is defined by these two transverse fields, and *not* by the electric and magnetic fields $\vec{E}$ and $\vec{H}$. In commonly-available materials and at optical frequencies, $\vec{B}$ and $\vec{H}$ are related by the scalar μ_0 and are therefore parallel, but in crystals the fields $\vec{D}$ and $\vec{E}$ are related by the dielectric tensor $\boldsymbol{\varepsilon}$, and are therefore not generally parallel; this is the background of some of the experiments suggested below [1]. In recent years there has been a lot of interest in synthetic meta-materials in which both $\boldsymbol{\varepsilon}$ and $\boldsymbol{\mu}$ can be tensors and even have negative values [2].

3.1.1 Ordinary and extraordinary light rays in crystals

When light energy travels through a medium, it follows a route called a *ray*, which is the locus of maximum intensity. In terms of Maxwell's theory, this ray has the direction of the Poynting vector $\vec{S} = \vec{E} \times \vec{H}$. Since the propagation vector $\vec{k}$ is normal to $\vec{D}$ and $\vec{B}$, it follows that in a crystal, where $\vec{E}$ and $\vec{D}$ are not necessarily parallel, $\vec{S}$ and $\vec{k}$ are not necessarily parallel, and so *the light intensity does not follow the wave propagation vector*, which is defined by Snell's law. In general, we define two types of light ray in a crystal: *ordinary rays*, whose ray vectors are parallel to the wave vectors (figure 3.1(a)), and *extraordinary rays*, where there is an angle

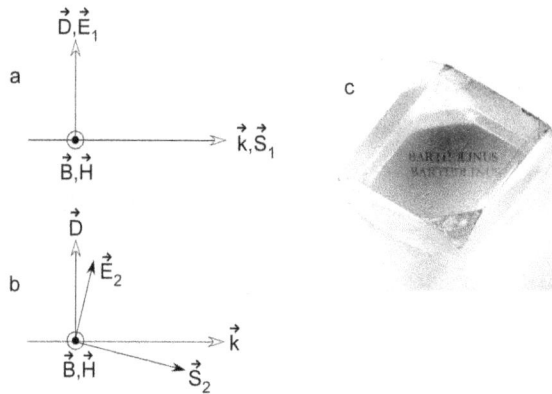

Figure 3.1. (a) Ordinary ray, for which the Poynting (ray) vector $\vec{S_1} \equiv \vec{E} \times \vec{H}$ is parallel to $\vec{k}$, (b) extraordinary ray where $\vec{S_2}$ is not parallel to $\vec{k}$, (c) image using unpolarized light through a birefringent calcite crystal, showing images due to both rays. Reproduced from [1] with permission of Cambridge University Press © 2011.

between the two (figure 3.1(b)). This sounds quite confusing, but can be seen quite clearly when looking at an object through a birefringent crystal, where two displaced images are formed (figure 3.1(c)). The fact that these two images are formed led to the name 'birefringent' for crystals having optical anisotropy. The most common birefringent crystals are *uniaxial* crystals and have one ordinary and one extraordinary ray in each $\vec{k}$-direction; the two coincide along a specific optic axis and are both ordinary. There are also *biaxial* crystals in which most $\vec{k}$-directions have two extraordinary rays. These crystals are discussed in more detail in section 7.2. It is important to notice that since the $\vec{k}$-vectors always obey Snell's law at an interface, the ray vectors of an extraordinary ray at the interface will not do so. In particular, an incident ray in air which is normal to the surface of a calcite crystal can create an extraordinary ray inside the crystal which is definitely inclined to the normal.

3.1.2 Types of polarized light

There are two basic types of polarized light: *linearly polarized* (also called *plane-polarized)* and *circularly* or *elliptically polarized.* In linearly-polarized light, the directions of the field vectors are given; for example a wave propagating with $\vec{k}$ parallel to $\hat{z}$ in a particular isotropic medium might have the field $\vec{D}$ in the $\hat{x}$-direction. An independent orthogonally-polarized wave travelling along the same direction would then have the field $\vec{D}$ in the $\hat{y}$-direction. A wave with a field in any other direction in the (x–y) plane can be described as a superposition of these two, when they have the same phase. Now a circularly-polarized wave in the same medium has a $\vec{D}$-field which rotates around the $\hat{z}$-axis once per period or wavelength, which can be described as a superposition of these same two orthogonally-polarized waves with a phase difference of $\pm\pi/2$ between them. There are two independent circularly-polarized waves, corresponding to the positive and negative phase differences. One rotates clockwise (looking into the propagating beam) and is called a 'right-handed circularly-polarized wave' and the other left-handed polarized wave rotating anti-clockwise. From a basic point of view, the circularly-polarized wave description is probably more fundamental, since the right and left-handed senses of rotation can be related directly to the photon spins $\pm\hbar$. In elliptically-polarized light, the two superposed orthogonally-polarized waves have different amplitudes, so that the resultant wave vector changes in length as it rotates.

3.1.3 Creation of polarized light

Some light sources, such as lasers, may emit polarized light because of some anisotropy in their construction, but generally light sources (lasers, LEDs, discharge lamps, incandescent lamps) emit unpolarized light, which is a random superposition of polarizations. There are several ways of creating polarized beams from an unpolarized source.

(a) *Linearly-polarized light.* The most common method of creating a linearly-polarized light beam is to pass an unpolarized beam through a dichroic material, which selectively absorbs one linear polarization. Commonly available 'Polaroid' sheet is one such material. This is basically a polymer sheet in which conducting polymer chains have been aligned by stretching the sheet. When the polarization of incident light is such that its electric field component is parallel to the chains, it then creates in them an electric current which is resistively dissipated and so the material absorbs that polarization. On the other hand, the orthogonal polarization, normal to the chains, creates no current and is not absorbed[1].

Another method of polarizing light from unpolarized sources uses reflection at the Brewster angle of incidence, which selects the wave with electric field normal to the plane of incidence (section 3.2.1). Brewster discovered experimentally that when a light ray is reflected from the surface of a transparent medium at an angle $= \arctan(n)$, where n is the refractive index, the ray is polarized normal to the plane of incidence. This can be proved from the Fresnel equations for reflection and transmission at an interface, but Brewster's discovery was essentially empirical.

Other widely used methods of polarization such as Nicol and Glan prisms are based on birefringent crystals. These crystals are discussed in section 7.2. The general idea is that an anisotropic crystal has a dielectric tensor which expresses the linear relationship between the $\vec{D}$ and $\vec{E}$ fields. The tensor has three mutually orthogonal principal axes. Referred to these axes ($i, j, k = 1, 2$ or 3) the vectors $\vec{D_i}$ and $\vec{E_i}$ are parallel to each other, but each one has a specific principal dielectric constant ϵ_i. A wave travelling along the i-axis can have two possible refractive indices. Assuming magnetic isotropy, the wave which has fields $\vec{D_j}$ and $\vec{B_k}$ then has refractive index

---

[1] This process of polarization can be vividly demonstrated using 3 cm microwaves, for which a grid polarizer can be constructed using parallel copper wires spaced by about 8 mm, which makes it obvious how it works.

$n_j = \sqrt{\epsilon_j}$, whereas the wave which has fields $\overrightarrow{D_k}$ and $\overrightarrow{B_j}$ has refractive index $n_k = \sqrt{\epsilon_k}$. Thus such a material is called 'birefringent' (figure 3.1).

The Nicol and Glan prisms are based on the idea that since the two refractive indices are different, the critical angle of reflection at an interface with an isotropic medium or between two crystals of the same material but having different orientations, is polarization-dependent. As a result, an unpolarized wave (which is a mixture of both polarizations) can be separated into its two components when the angle of incidence at the interface is between the two critical angles. Two such prisms are shown in figure 3.2

After the direction of the polarized field has been selected, it can be rotated without losing energy by inserting a half-wave plate, which is like the quarter-wave plate described below but has twice its thickness; the plate changes the polarization by geometrical reflection in its principal axes.

(b) *Circularly- and elliptically-polarized monochromatic light.* Circularly-polarized light is created from linearly-polarized light by using a *quarter-wave plate,* which is an anisotropic plate made from material (often from mica) with different refractive indices n_x and n_y for x- and y-polarized waves. If the thickness of the plate d is such that $(n_x - n_y)d = \pm\lambda/4$, then a phase difference of $\pm\pi/2$ is created between waves with x- and y-polarizations. Note that it is an advantage if $n_x - n_y$ is small, as in mica ($n_x = 1.596$, $n_y = 1.601$, $n_z = 1.563$ at wavelength 590 nm), so that the plate is fairly thick. The axis with the larger refractive index is called the 'slow' axis and the normal to it is the 'fast' axis. If a wave polarized at $\pm45°$ to these axes is incident on the plate it emerges circularly polarized, one sign giving left-handed and the other right-handed polarization. If the angle is not exactly $\pm45°$, the emergent wave is elliptically-polarized. Clearly, such a device is wavelength-dependent.

3.1.4 Characterizing the polarizers

It is often important to determine how close is a polarizer to an ideal device, which creates a perfectly polarized wave. This can easily be visualized with two identical

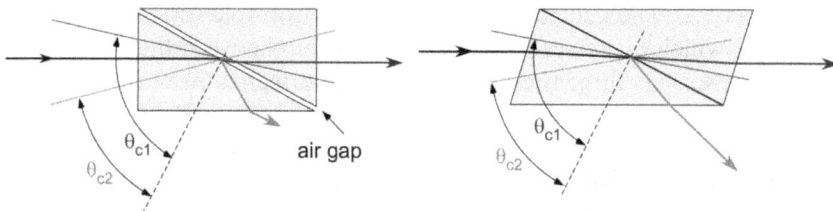

Figure 3.2. (a) Glan–air prism and (b) Nicol prism. In the former, the two prisms have parallel principal axes, and are separated by a thin air-gap. In the latter, the prisms have orthogonal axes and are cemented together. In both figures, the two orthogonal polarizations are indicated by red and blue colours.

polarizers, by superimposing them at right angles ('crossed'). For polaroid, the transmitted intensity of red light is found to be about 1% of that when they are parallel, and larger at the blue end of the spectrum. In more detail, the angle θ between the polarizers can be varied systematically and then the intensity should vary as $\cos^2 \theta$, i.e. $I(\theta) = a \cos^2 \theta + b$ (Malus's law). The extinction ratio, which is the ratio between intensities observed through crossed and parallel polarizers, is $b/(a + b)$. Note that this law is so basic that any deviation from it is likely to be due to inaccuracy in the calibration of the detector used (see section 1.2). Polarizing devices using birefringent crystals have much smaller (better) extinction ratios than polaroid film.

3.2 Fresnel coefficients for reflection at an interface

3.2.1 Fresnel coefficients

The Fresnel coefficients [3] describe the amplitudes of reflection and transmission of an incident wave at the interface between two dielectric materials. They were originally calculated by Fresnel based on a mechanical model for transverse waves, but were later justified for electromagnetic waves as well. They also apply to materials with generally complex dielectric constants. The wave is incident at angle θ_1 to the interface normal in a material with refractive index n_1, and is refracted to θ_2 in medium n_2. It can have p-polarization, also called parallel or TM, in which the electric field E lies in the plane of incidence or it can have s-polarization, also called perpendicular or TE, in which the field is normal to that plane. The reflection coefficients have the form:

$$R_p = \frac{n_1 \cos \theta_2 - n_2 \cos \theta_1}{n_1 \cos \theta_2 + n_2 \cos \theta_1}; \ R_s = \frac{n_1 \cos \theta_1 - n_2 \cos \theta_2}{n_1 \cos \theta_1 + n_2 \cos \theta_2}$$

which are shown in figure 3.3. Important points to notice are:
1. there is no difference between R_p and R_s at normal incidence,
2. $R_p = 0$ at the Brewster angle θ_B given by $\tan \theta_B = \cot \theta_2 = n_2/n_1$,
3. the phase of the reflected wave changes by π for the p-wave at the Brewster angle,

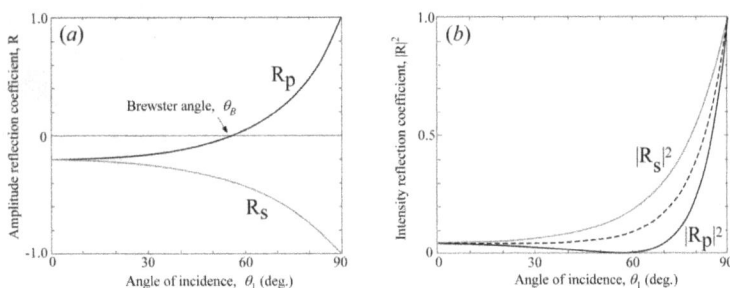

Figure 3.3. Fresnel coefficients for $n_2/n_1 = 1.5$. (a) amplitudes, (b) intensities.

4. at glancing angle in the less-dense medium the reflection coefficients are ±1,
5. above the critical angle in the denser medium both coefficients have unit modulus, but different phases (section 3.2.3).

3.2.2 Measuring the Fresnel coefficients

The measurement of $|R|^2$ requires a long sample with a plane surface, a stable light source and a photoelectric light-measuring detector. The source can be a laser or a wide-band source, because the dependence of n on wavelength is too small to be noticed in this experiment; the temporal stability of the light source and its polarization are important.

The set-up is quite simple, but it is important to make sure that incidence angles as close as possible to both 0° and 90° are accessible (figure 3.4). A long sample is required because close to 90° even a narrow beam of diameter w projects into an ellipse of length $w/\cos\theta \rightarrow \infty$ on the surface. The thickness of the sample is also a consideration. Ideally, one would like to have a single interface. This can be obtained by using a prism (check that it is uncoated!) and ignoring the refracted beam; an ideal sample is a long prism mounted horizontally (inset to figure 3.4), but this may not be available. With a laser source, the incident beam may be 1 mm in diameter. At grazing angle, say 87°, the beam width in the incidence plane is 20 mm; allowing for inaccuracies in set-up (such as in aligning the axis of rotation to lie in the sample surface), this would suggest using a 35 mm right-angled prism with a 50 mm hypotenuse. The sample surface must be clean. The incident light beam should be collimated and correctly polarized; the latter can be done by identifying the Brewster angle, where the reflected intensity should be zero, and adjusting the polarizer for minimum intensity, which identifies the p polarization. The s polarization is then obtained by rotating the polarizer axis by 90°. Since a laser source

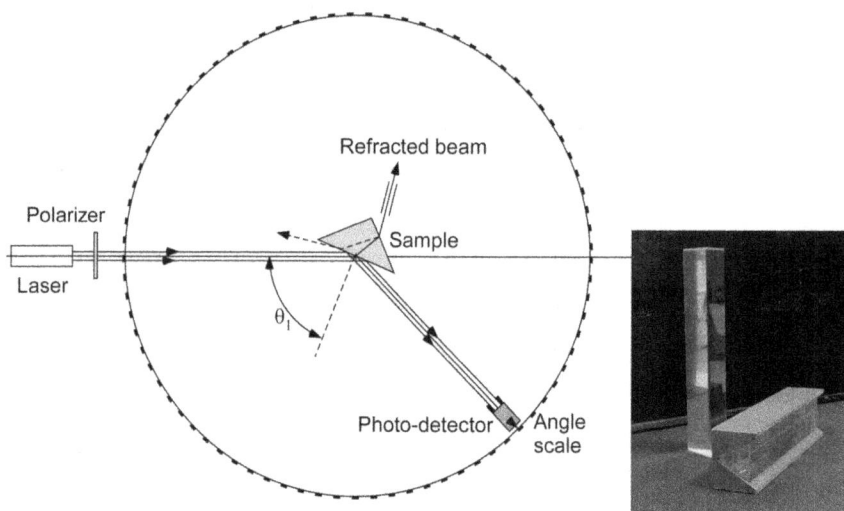

Figure 3.4. Setup for measuring external Fresnel coefficients using a laser source. The inset shows a long prism which is an ideal sample for this experiment, and a suggested mounting for it.

may have an intrinsic polarization, either the experimental results should be normalized by requiring that points (1) and (4) above are satisfied, or a quarter-wave plate can be inserted to circularly polarize the laser light, and the angle of its axes adjusted till the output intensity measured after the polarizer, before the sample is inserted, is independent of the polarizer orientation.

However, a suitable uncoated prism of a material other than glass may not be available. If a parallel-sided plate is used, the reflection from the back side (and multiple reflections) must be eliminated or corrected for; since the various reflections are laterally displaced, it is not too difficult to separate the front reflection by using an aperture before the photo-detector. However, this cannot be done at small angles of incidence. Another approach would be to use a very thin sample and collect all the reflections, but this introduces problems of interference if the light is coherent. The best solution seems to be the use of a sample which is as thick as possible, and to separate the reflected beams. If the laser beam is 1 mm diameter and the sample 1 cm thick, there is no great problem in separating the reflected beams down to an incidence angle of about 5°.

Two ways of getting quantitative results which can be compared to the theoretical values should be considered. The first is to measure the values of I_p and I_s as a function of θ, and normalize correctly. The second is to set the incident polarization at 45° and then to measure the polarization angle of the reflected beam by finding its extinction angle. This angle is $\arctan(R_p/R_s)$, which preserves information on the sign of R_p when going through θ_B, but does not give independent values for the two coefficients.

3.2.3 Incidence within the medium

If a suitable sample can be obtained, it is interesting to repeat the experiment with the light incident within the medium. Such a sample is a semi-cylinder (figure 3.5),

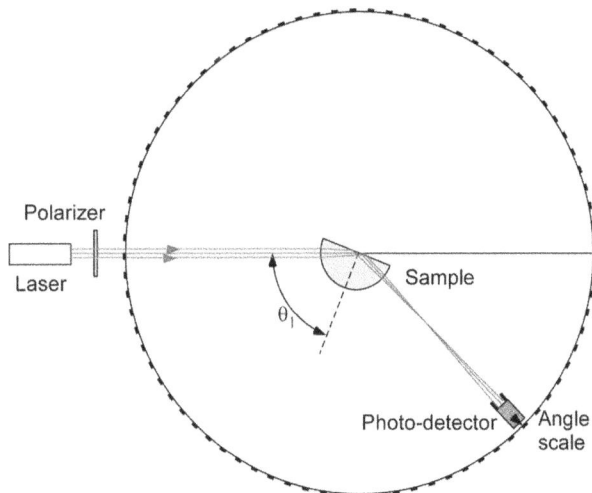

Figure 3.5. Using a semi-cylindrical sample to investigate incidence from within the medium.

when the axis of rotation now coincides with the axis of the cylinder. The angle of incidence can be varied and the phenomena of total internal reflection and critical angle explored. Above the critical angle, $|R| = 1$ but a phase change is introduced on reflection, which is different for the two polarizations (figure 3.6). How can the phase difference between the s- and p-polarized reflected waves be measured? One way could be measurement of the degree of elliptical polarization of the reflected wave.

3.2.4 Using total internal reflection to create circularly-polarized polychromatic light: Fresnel Rhomb

The Fresnel rhomb employs the phase change α introduced when light is reflected at angle $\hat{i}$ above the critical angle (total internal reflection), which is different for the parallel and perpendicular polarizations (figure 3.6):

$$\alpha_s = 2 \tan^{-1} \left(\frac{n \cos(\hat{i})}{\sqrt{n^2 \sin^2(\hat{i}) - 1}} \right)$$

$$\alpha_p = 2 \tan^{-1} \left(\frac{\cos(\hat{i})}{n\sqrt{n^2 \sin^2(\hat{i}) - 1}} \right).$$

It appears that for glass, with refractive index 1.5, the value of $\alpha_s - \alpha_p$ peaks at $45°$ when the angle of incidence is $52°$, so that two successive total reflections at this incidence result in a phase difference of $90°$, thus converting a wave which is linearly

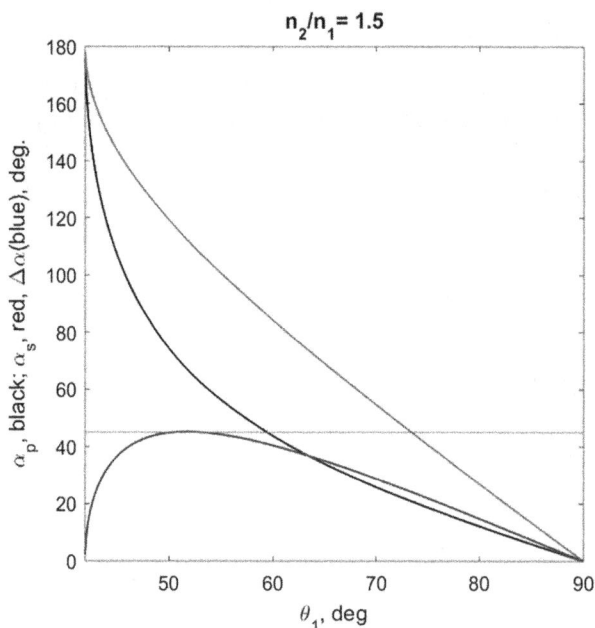

Figure 3.6. The phases α of critically reflected s (red) and p (black) polarized waves for $n = 1.5$, as a function of angle of incidence from within the glass, and their difference $\alpha_s - \alpha_p$.

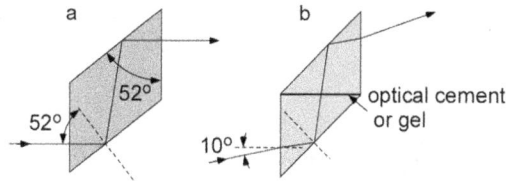

Figure 3.7. (a) Fresnel rhomb of glass with $n = 1.5$: the input wave is linearly polarized at 45° to the plane of incidence, and the output is then circularly polarized. (a) Optimal design and (b) an approximate construction from two 45° prisms cemented together.

polarized at 45° to the incidence plane to a circularly polarized wave. The result is almost achromatic, to the extent that the dispersion of the refractive index of glass is quite small. This system can be approximated by using two 45° prisms cemented together, with an angle of incidence of 10° to the entry plane normal (figure 3.7).

3.3 Ellipsometry: using polarized light to measure properties of thin films

3.3.1 The basic ellipsometer layout

The ellipsometer [4, 5] is an important instrument which is widely used to determine the properties of single or multiple thin films deposited on a plane substrate. The basic idea is to measure the reflection coefficients of light reflected from the object for various angles of incidence and polarizations, and to determine the required properties by comparison with a model, which is based on the thicknesses and complex refractive indices of the various layers involved, the angle of incidence, wavelength and polarization of the light. The commonest example is a thin layer of silicon dioxide deposited on a silicon substrate. In various experiments, the method has been shown to be capable of measuring thickness to an accuracy of better than one nanometer. Instruments which are automatically controlled by a computer are commercially available.

Various methods of arranging the optics are possible, but we will limit our discussion to the simplest example, which is a nulling-ellipsometer used to determine the thickness and refractive index of a single non-absorbing layer. The layout is shown in figure 3.8. The layout of commercial instruments is usually in a vertical plane so that samples, including liquids, can be most easily inserted.

The laser should be unpolarized to achieve most reliable results; otherwise if the polarizer is normal to the laser polarization, there will always be a null! If there is any uncertainty about the laser polarization, inserting a quarter-wave plate before the polarizer will solve the problem; it should be rotated until the polarizer passes light with more-or-less equal intensity for all orientations. This is sufficient for a null ellipsometer. The quarter-wave plate shown in the figure (QWP) is set with its axes at 45° to the x–z plane, and the polarizer and analyzer can be rotated to angles P and A with respect to this plane, where positive P and A are rotations anti-clockwise if you were looking into the beam. The angle θ can be varied, but most sensitivity is

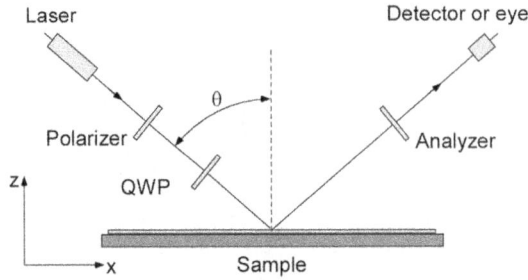

Figure 3.8. Ellipsometer layout.

obtained near the Brewster angles of the materials. Note that it is important to check that the polarizer and analyzer rotation scale zeros correspond exactly to polarization parallel to the x–z plane (the p-axis).

3.3.2 Samples

It is often possible to obtain samples of silicon wafers with a thin transparent layer of SiO_2 evaporated or sputtered onto it, or a polymer layer spun onto it. Another easily available sample is a glass prism or window with an anti-reflection coating on it, usually a quarter-wave layer of MgF_2. Since such an anti-reflection coating would be intended for green light at normal incidence, sufficient light from a He–Ne laser is reflected at oblique incidence to make the experiment possible. If the optics are in the vertical plane and the sample is horizontal, adsorbed layers of oil or soap on water can also be investigated.

3.3.3 Measurement method

The measurement is made by comparing the data with a simulation of a model. The result of the simulation is expressed in terms of two parameters Ψ and Δ, which are defined in terms of the complex reflection coefficients $r_p(\theta)$ and $r_s(\theta)$ for the p and s polarizations, as:

$$\tan \Psi \exp i\Delta = r_p/r_s.$$

For example, for a single thin film of thickness d and refractive index n_1 on a substrate with index n_2, with incidence in air with index $n_0 = 1$, we have first the Fresnel reflection coefficients at the upper (01) and lower (12) surfaces [3]:

$$r_{01,p} = \frac{n_1 \cos \theta_0 - n_0 \cos \theta_1}{n_1 \cos \theta_0 + n_0 \cos \theta_1}; \; r_{01,s} = \frac{n_0 \cos \theta_0 - n_1 \cos \theta_1}{n_0 \cos \theta_0 + n_1 \cos \theta_1}$$

$$r_{12,p} = \frac{n_2 \cos \theta_1 - n_1 \cos \theta_2}{n_2 \cos \theta_1 + n_1 \cos \theta_2}; \; r_{12,s} = \frac{n_1 \cos \theta_1 - n_2 \cos \theta_2}{n_1 \cos \theta_1 + n_2 \cos \theta_2}$$

in which the angles θ are given by Snell's law:

$$n_0 \sin \theta_0 = n_1 \sin \theta_1 = n_2 \sin \theta_2 = \sin \theta.$$

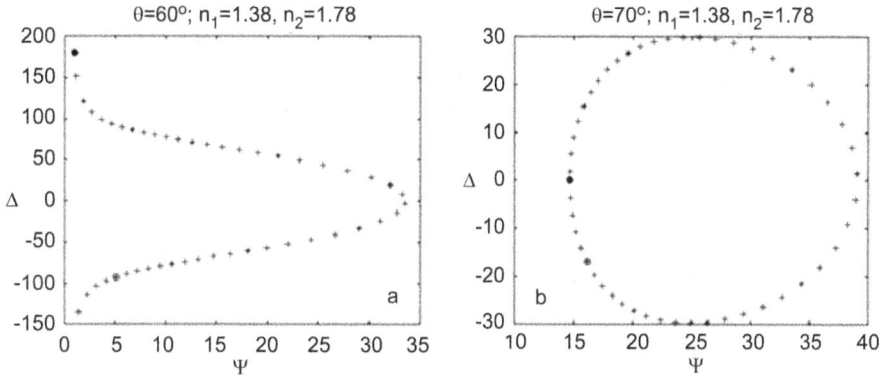

Figure 3.9. ($\Psi - \Delta$) curves for a layer of MgF$_2$ on SF11 glass at angles of incidence 60° and 70°, at thickness intervals of 0.02λ. The zero thickness is on the axis shown by a black spot, and 0.1λ by a red circle (to indicate in which direction the thickness increases). Notice that on the left-hand graph Δ jumps 360° at a thickness of d for which $2n_1 d \cos \theta_1 = \lambda$. The change in form occurs when the incidence angle goes through the Brewster angle of the substrate (61°).

Then multiple-beam (Fabry–Perot, section 5.4) interference between the two surfaces gives for each polarization (see appendix 1, section 3.3.4)

$$r = \frac{r_{01} + r_{12} \exp i\phi}{1 + r_{01} r_{12} \exp i\phi}; \quad \phi = 2n_1 k_0 d \cos \theta_1$$

from which tan Ψ and Δ can be calculated. Note that ϕ is the phase difference between successive reflected waves. Two examples of the output are shown in figure 3.9 for a film of MgF$_2$ on SF11 glass. These materials have refractive indices 1.38 and 1.78, respectively, and the two diagrams are for different incidence angles.

The experiment consists of finding, for a given value of θ, the angles P and A which result in complete extinction of the detected light. There are in general two such pairs of angles (see section 3.3.5, appendix 2):

1. When $-45° < P < 135°$, and $0° < A < 90°$, then $\Delta = 270° - 2P$ and $\Psi = A$,
2. When $-135° < P < 45°$, and $-90° < A < 0°$, then $\Delta = 90° - 2P$ and $\Psi = -A$.

Measuring both values for a given value of θ gives an indication of the accuracy to which the QWP was set to 45°. The experiment should be repeated at several values of θ, for each of which the simulation has to be carried out. Each value of θ gives you an estimate of the thickness d. Of course, in practice the measured values of Ψ and Δ may not lie exactly on the theoretical curves; the error might be statistical (noise) or it might be systematic, because one of the values of refractive index was not correct. In that case, changing the values in the simulation might correct the error. If fact, you can use this way to measure the index of the film, which might be different from the bulk material if the film is very thin or has different crystalline order from the bulk.

We have sometimes found that there are serious discrepancies between the measured data and its simulation, in which case the following points should be checked. First, check that the rotation angles of the polarizers are measured clockwise from the vertical when looking into the beam (figuratively, of course). That means that the polarizer should be rotating clockwise when you look through it into the laser, and the analyzer when you look back through it in the direction of the sample. Second you should check for mistakes in the simulation program by using it to simulate cases where the answer is well known. For example, if the film has zero thickness, the values of Δ and Ψ should correspond to the reflection coefficients (figure 3.3) of a single interface between air and the substrate, including the phase jump of $180°$ at Brewster's angle.

3.3.4 Appendix 1: Derivation of the multiple reflection amplitude

Derivation of the basic formula for the amplitude reflection coefficient, which looks very simple, is tricky and it is very easy to make mistakes in the signs of the expressions involved. First, in the case of the s-polarization, the amplitude of the electric field on the two sides of the boundary between any pair of dielectric materials is continuous, so that we have $1 + r_{ij} = t_{ij}$, where the incident amplitude is 1 and r and t are the reflected field and transmitted field amplitudes respectively. In the case of the p-polarization, the same is true of the magnetic field. Now, we use the formulae for r_{ij} at each pair of interfaces to calculate the total reflected amplitudes r_p and r_s from the film defined by subscripts 0 for the incident medium, usually air or vacuum, 1 for the film and 2 for the substrate, where the phase delay introduced between successive reflections which exit the film is $\phi = 2n_1 dk_0 \cos\theta_1$:

$$r = r_{01} + \left\{ t_{01}r_{12}t_{10}\text{exp}i\phi + t_{01}r_{12}r_{10}r_{12}t_{10}\text{exp}2i\phi + t_{01}r_{12}(r_{10}r_{12})^2 t_{10}\text{exp}3i\phi + \ldots \right\}$$

In the above equation the r_{ij}'s and t_{ij}'s are written in the order which they occur in each term. Summing the geometrical series we get

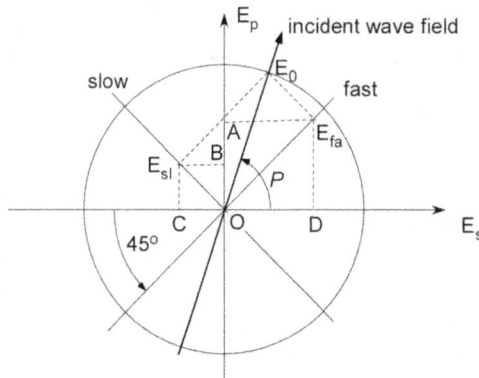

Figure 3.10. Electric field vectors in the QWP plane.

$$r = r_{01} + \frac{t_{01}t_{10}r_{12}\exp i\phi}{1 - r_{10}r_{12}\exp i\phi} = \frac{r_{01}(1 - r_{10}r_{12}\exp i\phi) + t_{01}t_{10}r_{12}\exp i\phi}{1 - r_{10}r_{12}\exp i\phi}$$

This simplifies to

$$r = \frac{r_{01} - r_{12}(r_{01}r_{10} - t_{01}t_{10})\exp i\phi}{1 - r_{10}r_{12}\exp i\phi}$$

Now, $t_{ij} = 1 + r_{ij}$, so that $t_{ij}t_{ji} = (1 + r_{ij})(1 + r_{ji}) = 1 + r_{ij}r_{ji}$ since $r_{ij} = -r_{ji}$, so that finally

$$r = \frac{r_{01} + r_{12}\exp i\phi}{1 - r_{10}r_{12}\exp i\phi}$$

which is the expression on which single-film ellipsometry is based.

3.3.5 Appendix 2: Derivation of the null angles

The QWP has its fast and slow axes at angles 45° and −45° to the sample normal. The angle of the polarizer is P. Then, referring to figure 3.10, the components of the incident wave on the fast and slow axes are:

$$E_{fa} = E_0 \cos(P - 45°); \quad E_{sl} = E_0 \sin(P - 45°).$$

Since the effect of the QWP is to introduce a phase change of $\pi/2$ between the above fields, the s- and p-polarized fields incident on the sample are

$$E_p = OA + OB$$
$$= E_0 \cos 45°[E_{fa} + iE_{sl}]$$
$$= E_0 \cos 45°[\cos(P - 45°) + i \sin(P - 45°)]$$
$$= E_0 \cos 45°\exp[i(P - 45°)]$$

$$E_s = OD - OC$$
$$= E_0 \cos 45°[E_{fa} - iE_{sl}]$$
$$= E_0 \cos 45°[\cos(P - 45°) - i \sin(P - 45°)]$$
$$= E_0 \cos 45°\exp[-i(P - 45°)].$$

Note that $|E_p| = |E_s|$. The phase difference between E_p and E_s is thus $2(P - 45°)$. This cancels the phase difference Δ introduced by the sample when $\Delta = -2(P - 45°)$, i.e.

$$\Delta = 90° - 2P.$$

Now the angle A of the analyzer has to be adjusted to null the ratio between the amplitudes of the reflection coefficients r_p and r_s.

The reflected waves have amplitudes $E_{rp} = E_p r_p$ and $E_{rs} = E_s r_s$. The components of these along the analyzer give

$$E_A = E_{rp} \sin A + E_{rs} \cos A$$

Figure 3.11. A morning view of distant mountains (about 100 km to the north) in clear weather, photographed through horizontal (left) and vertical (right) polarizers.

which is zero when

$$\frac{\sin A}{\cos A} = -\frac{E_{rs}}{E_{rp}} = -\frac{r_s}{r_p} = -\tan \Psi$$

$$\Psi = -A.$$

Now show by the same method that there is a second nulling solution given by $\Delta = 270° - 2P$, $\Psi = A$.

3.4 Rayleigh scattering

3.4.1 Scattering of polarized light, photographic applications

When light is scattered by particles, the intensity of the scattered light depends on the polarization, particularly if the particles have dimensions of the order of, or smaller than, the light wavelength [6]. The best-known and simplest example is 'Rayleigh scattering' which applies when the size is considerably less than the wavelength. Then, the scatterers behave like dipoles oscillating in the direction of the electric field of the incident wave, and their scattering strengths are proportional to λ^{-4}, which is why the sky is blue. It is easy to see that when the scattered light is at 90° to the incident ray and the polarization is in the plane of incidence the scattering is weakest. If the incident ray is unpolarized, the scattered light is thus polarized normal to the plane of incidence; it can thus be eliminated by observing through a polarizer oriented in the plane. This effect is often used by photographers to substantially reduce scattering of sunlight by air and haze, which reduce the contrast when photographing distant scenes (figure 3.11).

Experiments on Rayleigh scattering in the lab can be done using a medium of *very* dilute milk, in which the oil drops are very small and the probability of light being scattered successively by two or more particles is negligibly small. For example, pass an unexpanded polarized laser beam along the axis of a tube containing the solution and measure the dependence of the scattered intensity on the angle between the incident light and observer, and its dependence on the laser's plane of polarization. The theory of scattering by larger particles is quite complicated [5].

3.4.2 Wavelength dependence of Rayleigh scattering

The dependence of Rayleigh scattering on the wavelength can be investigated in two ways. First, using air as the scattering medium, a series of photographs such as figure 3.11(left) can be taken through colour filters at known wavelengths. It is important that the photographed region includes clear blue sky and some white scattering surfaces. The intensity of the light scattered by the sky in a given region can then be compared quantitatively to that scattered by the white surface, and the wavelength dependence deduced. Another way would be to use a spectrometer to compare the spectra of the light from a broad-band source scattered from the dilute milk to that from the same source scattered by a Lambertian white surface, such as white chalk or paper painted with white erasing fluid (TiO_2).

3.5 Coherent back-scattering

3.5.1 Localization of light by non-absorbing random materials

When a collimated light beam is incident on a medium which scatters it, but does not absorb it, the light appears in a cone whose angular size depends on the particular properties of the scattering elements. In general, the wave is returned after several scattering events, and its phase is uncorrelated with other scattered waves, so that the intensity of the light received at a given angle is the sum of the intensities of the individual scattered waves. However, if the received wave is at 180° angle to the incident, i.e. back-scattered, there is always a second identical multiple scattering event in which the wave travels the same route in the opposite direction and therefore has the same optical path length. The second exiting wave therefore has the same phase as the first and interference between the two is constructive. As a result, the back-scattered wave is twice as intense as the scattered waves at other angles. This is also true for waves returned within a small angular cone around the back-scattering direction, whose angle is determined by the density of scattering particles in the medium.

As a result of the above situation, if we look at the intensity of the light scattered back from such a medium, as a function of the angle of scattering α, we see a cone of light with a brighter spot on the axis at $\alpha = 180°$. The angular size of the spot, which is equal to the wavelength divided by the penetration depth of the waves in the medium, is a few mrad. In principle, the spot should be twice as intense as the neighbouring regions of the cone. If the scattered waves are coherent, their polarization should be preserved.

This effect, called 'coherent back-scattering' or 'weak localization' of light, was first observed in 1984 [7] in a medium consisting of sub-micron latex spheres suspended in water. The phenomenon was explained [8] and quantitatively confirmed [9, 10] the following year. Subsequently, the effect was demonstrated in other media, including milk and solids such as Teflon and white paper. A comprehensive treatment is given in reference [11].

3.5.2 Experiments

The experiments are quite simple, but require some attention to details. Most of the experiments published used illumination by a laser beam and detection by a photomultiplier masked by a pinhole and narrow-band filter. This detector was linearly translated to map the scattered light intensity quantitatively. In a different version using a camera (figure 3.12), we took a laser beam and expanded it to a plane wave of diameter of about 8 mm. This beam impinged on the sample after reflection by a beam-splitter. The back-scattered wave returned through the beam-splitter and continued to a lens with a camera in its focal plane, which recorded the angular dependence of the scattered intensity. Since the results depend on polarization, we placed a polarizer (P) in the incident beam and an analyzer (A) in the reflected beam. Several variations were investigated: both P and A vertical, both P and A horizontal, and P vertical with A horizontal. In addition, we tried an arrangement suggested by Corey *et al* [12] where a quarter-wave plate is added between the beam-splitter and the sample, so that an incident vertically polarized beam becomes circularly polarized and the scattered wave with the same sense is then converted to horizontally polarized, which passes through a horizontal analyzer; this should reject more singly-scattered light and give a stronger back-scattering peak.

Since a fraction of the incident light is scattered or reflected without reaching the sample, it is important that this light should not reach the camera, and suitable baffles were put in place. In particular, the (approximately) 50% of the incident beam transmitted by the beam-splitter must be effectively absorbed, since any of it which returns to the beam-splitter will be focused to a point exactly on the axis where the coherent back-scattering is observed. To obliterate this beam, we collected it in a beam dump at an appreciable distance from the beam-splitter. We

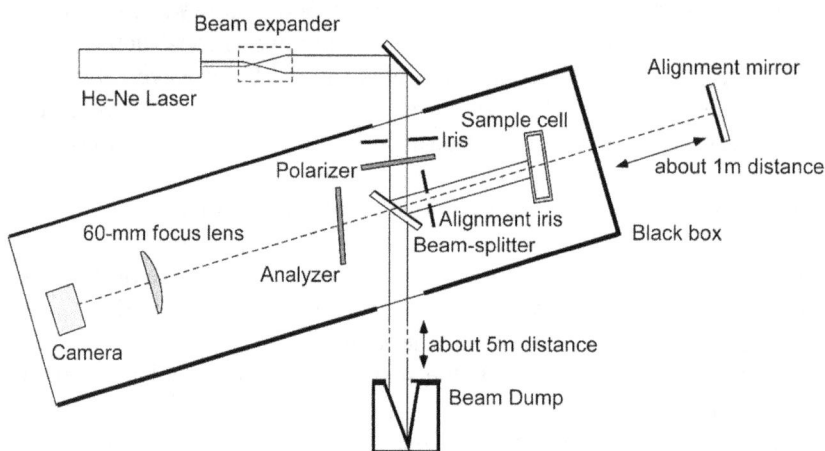

Figure 3.12. Experimental set-up used to observe coherent back-scattering. During initial alignment, the alignment iris was closed to about 2 mm diameter and the 'Black box' and sample cell were removed; the alignment mirror and the lens were then adjusted to give a sharp symmetrical peak at the camera. This defines the 180° back-scattering axis. Then the black box and sample cell were put back and the alignment iris was fully opened, before experimenting with a real sample.

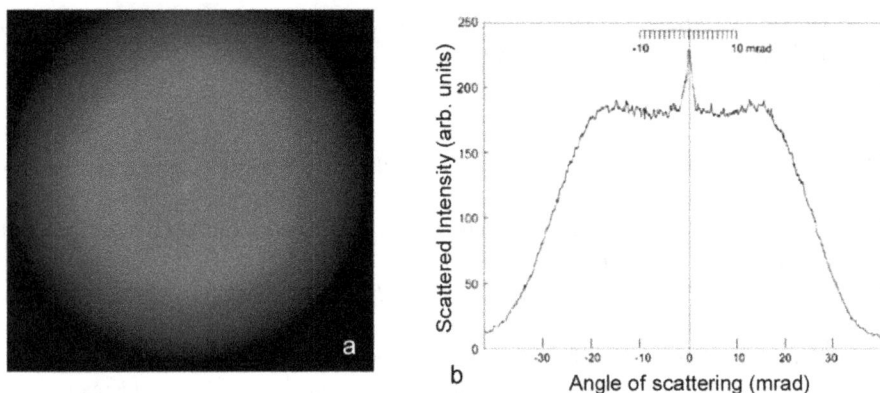

Figure 3.13. Experimental scattering intensity from 3% homogenized milk: (a) observed scattering, (b) intensity measured along a diametrical axis.

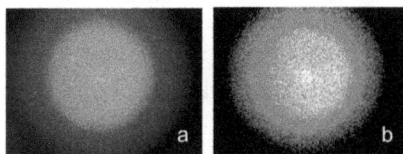

Figure 3.14. Experimental scattering intensity from white paper. (a) The scattering length is shorter, so the peak is weaker and its angular size is larger. (b) Contrast of (a) has been enhanced for visibility.

must also avoid any specular reflection by optical elements reaching the detector; the polarizers and the face of the cuvette containing a liquid sample should be inclined to the optical axis, and it is better to use a non-polarizing plate beam-splitter than a cube, since there are less normal surfaces in the former. To ensure that the camera was in the focal plane of the lens, the system was initially aligned with the sample replaced by an external plane mirror whose orientation was adjusted to return the light exactly through an iris situated on the optical axis (see caption to figure 3.12). We also confirmed, by covering the remote beam dump temporarily with white paper, that the image of any light scattered by it would appear exactly at the same focal point. For this reason, the blackness of the beam dump is critical.

An example of results using milk is shown in figure 3.13. White paper and other white solids also show the coherent back-scattering peak, but on a much noisier background (figure 3.14).

These experiments used a CCD camera, which should have linear response to intensity up to the saturation level. However, many cameras have CMOS or other sensors, which are not linear. To obtain quantitative results, the camera response should therefore be calibrated; a calibration procedure is described briefly in section 1.2. Since liquid samples such as milk are dynamic, it is an advantage to record the images with a long exposure time (of order 1 s) to reduce noise.

A further effect which is visible in the results is a weakening of the incoherent scattering at angles within a few degrees of the back-scattering peak (figure 3.13(b)).

This is a result of energy conservation (where does the energy of the peak come from?), and has recently been analyzed theoretically [13].

References

[1] Lipson A, Lipson S G and Lipson H 2011 *Optical Physics* 4th edn (Cambridge: Cambridge University Press) sections 6.5, 6.10

[2] Shelby R A, Smith D R and Shultz S 2001 Experimental verification of a negative index of refraction *Science* **292** 77–9

[3] Lipson A, Lipson S G and Lipson H 2011 *Optical Physics* 4th edn (Cambridge: Cambridge University Press) sections 5.4, 5.5

[4] The ellipsometer http://ecee.colorado.edu/~bart/book/ellipsom.htm

[5] Tompkins H G 1993 *A User's Guide to Ellipsometry* (San Diego, CA: Academic)

[6] van de Hulst H C 1981 *Light Scattering by Small Particles* (New York: Dover)

[7] Kuga Y and Ishimaru A 1984 Retroreflectance from a dense distribution of spherical particles *J. Opt. Soc. Am.* A **8** 831

[8] Akkermans E and Maynard R 1985 Weak localization of waves *J. Phys. Lett.* **46** 1045–53

[9] Van Albada M P and Lagendijk A 1985 Observation of weak localization of light in a random medium *Phys. Rev. Lett.* **55** 2692

[10] Wolf P-E and Maret G 1985 Weak localization and coherent backscattering of photons in disordered media *Phys. Rev. Lett.* **55** 2696

[11] Akkermans E and Montambaux G 2007 *Mesoscopic Physics of Electrons and Photons* (Cambridge: Cambridge University Press) ch 8

[12] Corey R, Kissner M and Saulnier P 1995 Coherent backscattering of light *Am. J. Phys.* **63** 560

[13] Fiebig S and Aegerter C M *et al* 2008 Conservation of energy in coherent backscattering of light *Eur. Phys. Lett.* **81** 64004

IOP Publishing

Optics Experiments and Demonstrations for Student Laboratories

Stephen G Lipson

Chapter 4

Physical optics I: diffraction and imaging

4.1 Fraunhofer (far-field) diffraction and Fourier transforms

This topic arose from Fraunhofer's analytical study of diffraction by gratings around 1821. It achieved great importance from the 1920s when it was realized by von Laue and the Braggs that crystals are essentially three-dimensional diffraction gratings, and that their structure can be elucidated by observing the diffraction patterns created using x-rays [1]. In this section, we shall limit the discussion to two-dimensional structures; in section 4.6 we present a simple three-dimensional example on the optical scale. The basic motif of Fraunhofer diffraction is that when a transmitting or reflecting mask is illuminated by a plane wave, the far-field diffraction field is the Fourier transform of the object transmission or scattering function and the observed pattern intensity is the square-modulus of that Fourier transform. The formalism is as follows for a transmitting mask in the plane $z = 0$, with complex amplitude transmission function $f(x,y)$, when it is illuminated by a coherent monochromatic plane wave travelling parallel to the z-axis [2]. In a distant plane at $z \to \infty$, a diffracted amplitude proportional to the transform $F(u, v)$ is observed, where F is the two-dimensional Fourier transform of $f(x,y)$, and $(u, v) \equiv \frac{2\pi}{\lambda}(\sin \theta_x, \sin \theta_y) \approx \frac{2\pi}{\lambda}(\theta_x, \theta_y)$ for small angles. The two-dimensional angular position in the distant plane is (θ_x, θ_y) relative to the z-axis. The observed intensity is proportional to $|F(u, v)|^2$. We can use a converging lens or lens combination with effective focal length F_{eff}, to translate these angles to positions (X, Y) in the lens's focal plane $z = Z_F$: $(X, Y) = F_{\text{eff}}(\theta_x, \theta_y)$ for small angles. The transmission function $f(x,y)$ is in general a complex function in which a change of phase in the transmitted wave is represented by the phase angle of the complex value, and a change in amplitude by its modulus.

doi:10.1088/978-0-7503-2300-0ch4   4-1

4.1.1 Optical setup

A very simple demonstration of Fraunhofer diffraction can be made with no hardware at all. Simply look at a point source of light (at any distance) and put a mask of suitable dimensions in front of your eye[1]. The mask should contain detail smaller than the eye's pupil. Centred on the image of the point source you can see the Fraunhofer pattern of the mask. If the source is monochromatic (for example a distant Na street lamp), the Fourier transform is demonstrated; if it is polychromatic, superposition of the patterns on different scales in different colours are seen. One example is a woven silk scarf, which is approximately a square lattice. Other masks could be small holes of different shapes, or a pair of slits about one mm apart.

In the laboratory, the various components of this experiment can be mounted on an optical rail. First, the monochromatic plane wave is created by using a laser, a beam expander and spatial filter, and a converging lens to create a plane wave of sufficient area to illuminate the object mask. The beam is not actually uniform in intensity, but has a Gaussian distribution with its maximum on the z-axis; this is important in some of the experiments (e.g. #11 below, Babinet's theorem). However, in most cases, it is best to have the extent of the Gaussian beam considerably larger than the size of the mask, so that the illumination can be considered uniform to about 10%. For the experiments on masks of macroscopic scale ($\gg\lambda$), the polarization of the beam is not relevant. The mask is placed in the plane wave, and is followed by the converging lens or lens combination with effective focal length F_{eff}. Although ideally the position of the mask should be in the front focal plane of the lens, an error in this position only affects the phase of the diffraction pattern, which is not observable, so that the mask can be placed as close as convenient to the lenses. The Fraunhofer pattern is then projected onto a camera sensor (CCD chip) or onto a semi-transparent screen (tracing paper) which can then be photographed from the other side using a camera on-axis. Essentially the CCD chip is better for work in which, say, the diffraction pattern is compared quantitatively to a calculated transform, but the latter is very convenient for students to record diffraction patterns on their own laptops or smart-phones. We will next calculate appropriate values of F_{eff} for these two configurations, and later describe how F_{eff} can conveniently be measured.

The value of a convenient effective focal length can be found as follows. Consider for example the diffraction pattern of a slit with width a. This is represented along the x-axis by a function $f(x) = \text{rect}(x/a)$, whose Fourier transform is $a\,\text{sinc}(ua/2)$. The transform has its first zeros at $u = \pm 2\pi/a$, which essentially defines the 'size' of details in the diffraction pattern to be demonstrated. Then the angular size will be $\theta_x = \lambda u/2\pi = \lambda/a$, which translates to $X = F_{\text{eff}}\theta_x = F_{\text{eff}}\lambda/a$ in the focal plane. Suppose $a = 2$ mm, which is a convenient size for manual construction. If we record the diffraction patterns on a CCD with pixel size of order 10 μm, for quantitative work we need this angle to be projected as about 50 pixels, i.e. $X = 0.5$ mm. This indicates that $F_{\text{eff}} \approx 2$ m for visible light. Such focal lengths can be achieved in a reasonable

---

[1] This is the condition $z = -z_1$ mentioned in section 4.2.

space using a telephoto combination of a positive and negative lens (e.g. +200 mm and −50 mm separated by 155 mm, which focuses at 450 mm from the second lens: figure 4.1(a)). On the other hand, if the diffraction patterns are to be projected onto a screen, the details should be at least 5 mm in extent, which requires a focal length of 20 m (figure 4.1(b)). Take into account the fact that the lenses must be accurately aligned and exactly normal to the axis for the best quality diffraction patterns. This should be checked by observing the diffraction pattern from a circular aperture with diameter equal to the largest masks anticipated; the pattern should clearly show the Airy-function rings around the central sharp spot. The effective focal length can be calibrated by using a periodic object of known or measured period (low frequency Ronchi ruling, for example) and measuring the distance between the orders.

4.1.2 Construction of diffraction objects

We are mainly concerned with masks which are sufficiently well-defined that their diffraction patterns can be calculated, either analytically or with a computer. Although masks can be purchased, we feel that it is important that students make and measure their masks themselves, and do not rely on some manufacturer's specifications. Amongst other things, this will encourage them to appreciate the results of errors in construction; for example, compare the diffraction patterns of a slit with parallel edges, and one with edges not quite parallel. We suggest three categories:

1. Masks which consist of regions which are either opaque or transparent at all points can be prepared from thick foil, opaque plastic film or good-quality cardboard which is either punched or drilled according to a prescription. For example, computer cards (do you remember them—the input to digital computers in the 1970s?) can often be found lying around, and provide nice cleanly cut rectangular holes and arrays of such holes. The good quality card

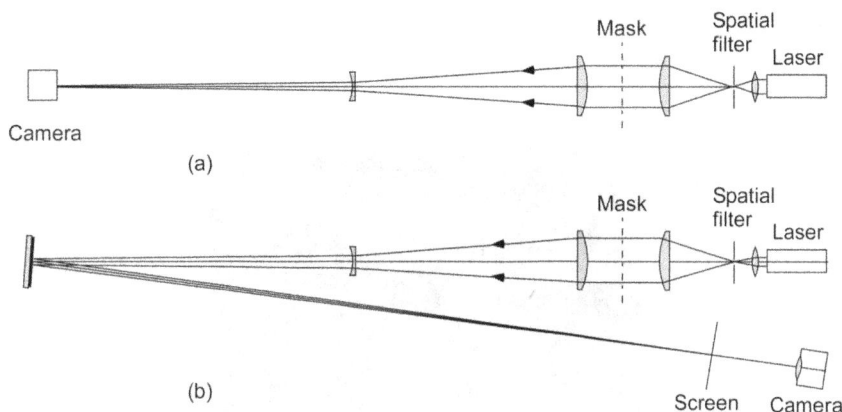

Figure 4.1. Setup for Fraunhofer diffraction using a telephoto combination to project the diffraction pattern (a) directly on the camera and (b) on a semi-transparent screen, which is photographed from the opposite side. In (b), the axis is folded by a plane mirror to save laboratory space and enlarge the pattern.

cuts with very clean edges, which is a must for diffraction masks, and a sharp knife can be used to cut shapes on the mm scale.

Masks of patterns consisting of round holes (e.g. figure 4.4) can be made using a drill press or a milling machine. An ideal, but rarely available, machine for preparing punched-hole masks is a pantograph (figure 4.2); a drawing of the required arrangement is prepared at about ten times the required size, and a cursor moved from point to point. The punch, which is used to make the holes, moves at about one-tenth of the drawing scale over the fixed card [3]. The mask of figure 4.6 was made this way. It is a nice mechanical engineering project to construct such a pantograph, and to check its reliability optically. However, for really accurate masks, machine-shop tools are needed.

Holes of non-circular shapes can often be made by preparing a punch of the right shape and using it on the card or film with a lead block underneath; for example, a hexagonal hole can be made by using a sharpened Allen key as the punch, or a triangular punch can be filed from a steel rod. One sharp tap with a hammer should cut a clean-edged hole. The accuracy of the work here can be quite important, as we'll see later.

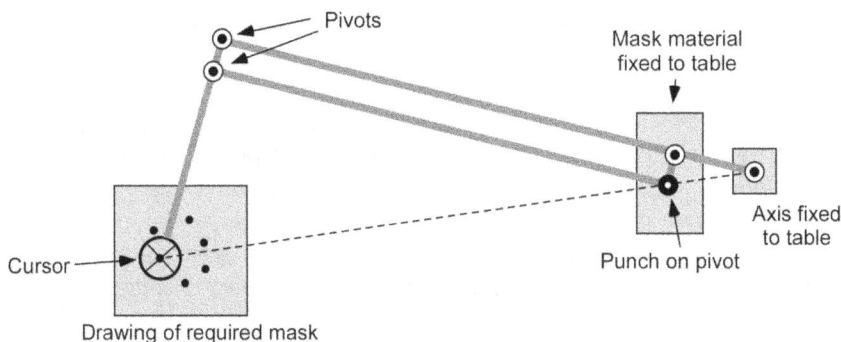

Figure 4.2. Schematic drawing of the pantograph. The cursor, punch and axis must lie on a straight line to create an undistorted copy.

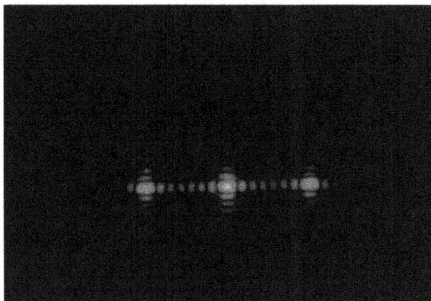

Figure 4.3. Diffraction by eight parallel slits. Reproduced from [2] with permission of Cambridge University Press © 2011.

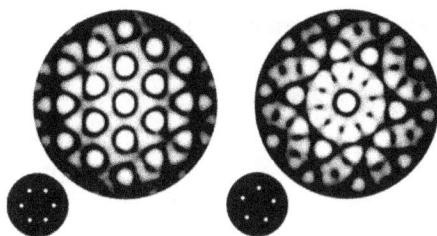

Figure 4.4. Six and five holes around a ring, and their diffraction patterns. Reproduced from [2] with permission of Cambridge University Press © 2011.

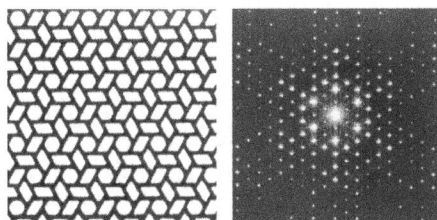

Figure 4.5. A Braun electric shaver part and its diffraction pattern. Reproduced from [2] with permission of Cambridge University Press © 2011.

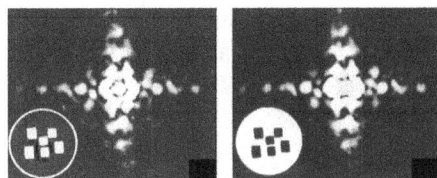

Figure 4.6. Babinet's theorem: diffraction patterns of a mask (shown in the circle at bottom left) and its complement. Reproduced from [2] with permission of Cambridge University Press © 2011.

2. Phase masks can be made by using the above techniques to produce an amplitude mask, and then covering some of the holes with thin glass (microscope cover slips). The phase shift obtained this way can be altered continuously by rotating the mask about an axis normal to the optical axis. Pieces of cleaved mica, which is atomically flat, might seem a better alternative, but mica is birefringent (section 3.1.3); unless the laser is polarized and the mica sheets are oriented consistently, the results may be problematic.

3. Masks with continuously varying transmission can be made photographically, although the equipment for this (film, developer, fixer) is becoming difficult to acquire. Complicated masks of category 1 above can also be made this way (e.g. a chessboard). There is a potential problem, that phase differences might be introduced by the developing process, but we have rarely found this to be serious.

4.1.3 15 ideas for significant diffraction objects

Now we can discuss some masks which are interesting and illustrate many of the principles of Fourier optics. Remember that what we photograph with the camera is the *square modulus of the Fourier transform* of the object mask, on a scale determined by the effective focal length of the diffraction lens optical system and the laser wavelength. A nice collection of such photographs has been published by Harburn *et al* [4]. We suggest that for every mask, the students should first study the principles to be demonstrated and should show *analytically* how the Fourier transform is related to the construction in terms of symmetry, dimensions, convolutions, products etc, including estimates of the relative intensity of important features. Only after this should a numerical Fourier transform be carried out, in order to show details which can be compared with the experimental results.

1. *Diffraction grating.* We mention this first because it provides a method of calibrating the system. A periodic mask of the form $f(x) = g(x) \otimes \sum_n \delta(x - nd)$ where d is the period, $g(x)$ is the line shape, $\otimes$ is the convolution operator and $\delta(x)$ is the Dirac δ-function, gives a pattern which is the set of diffraction orders m: $F(u) = \sum_m \delta(u - 2m\pi/d)$, multiplied by the intensity of the transform $|G(u)|^2$ of the shape function, $g(x)$. If we take a simple coarse grating, or a set of holes drilled at intervals of $d = 1$ mm for example, the diffraction orders m will be quite sharp and they can be used to calibrate the system since the distance between them is $2\pi/d$. In particular, when a telephoto combination is used, this obviates the necessity of knowing or measuring the effective focal length accurately. How does the fact that the mask has finite external dimensions affect the diffraction pattern?

2. *A slit aperture.* Since the experiment is two-dimensional, the slit must have both a height and a width; the diffraction pattern will therefore be that of a rectangular hole, but one can see that as the slit becomes longer, approaching an infinite length, the diffraction pattern contracts to the u-axis.

3. *A pair of slits or holes, and a phase object* (very conveniently taken from a punched computer card). This is basically Young's experiment. If one of the holes is covered by glass, the effect of a phase difference, which shifts the fringes along the u-axis, can be investigated qualitatively by rotating the mask about an axis normal to the optical axis of the setup. If the thickness of the glass and its refractive index are known, the experiment can be made quantitative by measuring fringe shift as a function of angular rotation α; the relationship is linear as a function of $\cos(\alpha)$ for small α.

4. *An arbitrary number of slits.* A periodic set of slits can be used, with a second variable-width single-slit mask in a plane close to it. The second mask is then used to select a given number of the periodic set for the experiment (figure 4.3). Ronchi rulings provide convenient periodic sets of slits, but their slit width is always equal to half the period, which is an undesired limitation.

5. *A generally-shaped polygonal aperture*, for example a triangle. One should emphasize here that the *intensity* of the diffraction pattern of any real function is centro-symmetric, even if the mask is not. Thus, the diffraction pattern of the triangle has six arms! The interpretation is considerably simplified if one looks at the contributions of the edges, each of which is locally a step-function. The modulus of the transform of the step-function is centro-symmetrical about the origin. What happens if there are two or more parallel edges to the polygon?

6. *Groups of similar apertures.* Interpret these in terms of convolution. If the arrangement is periodic, then we have a finite periodic crystal. A chessboard is a nice example.

7. *Round holes of various sizes.* One can also use a variable iris diaphragm, although this is in fact a many-sided polygon, which becomes clear from the diffraction pattern. If the mask is rotated about an axis normal to the optical axis of the setup, the projected hole becomes an ellipse. The diameter of the iris aperture is conveniently set by closing it on a calibrated circular rod or the shaft of a drill bit, which is then carefully removed.

8. *Group of n holes equally-spaced around a circle.* This is an important lesson in symmetry; when n is even, the diffraction pattern has n-fold symmetry, but when n is odd, the symmetry is $2n$-fold. This arises because we are looking at the intensity of the diffraction pattern, which does not distinguish between 0 and π phases. To get really good $2n$ symmetry in such diffraction patterns, the mask has to be made very accurately, because an error in the position of one hole results in additional two fold symmetry (figure 4.4). The radial dependence of the diffraction pattern is an nth order Bessel function.

9. *Lattices.* Superimposing two periodic coarse gratings (e.g. Ronchi rulings) having different periods at a variable angle gives a two-dimensional lattice. The diffraction pattern, which is the convolution of the two resulting strings of equally-spaced diffraction orders, is basic to understanding the reciprocal lattice underlying crystal diffraction. Some more complicated two-dimensional periodic gratings can be found ready-made, such as the electric shaver mask shown in figure 4.5. The diffraction pattern should be described as the product of the reciprocal lattice and the diffraction pattern of the unit cell detail. With patience, a mask can be drilled in which there are groups of holes (e.g. a pair) repeating on a lattice, or this can be produced photographically.

10. *Gauze; a periodic structure with errors.* Since mosquito-net gauze is not really accurate enough for quantitative diffraction experiments, it can be used to show the effects of errors in periodic structures. Moreover, because of the weaving process, there are weak 'superlattice' diffraction orders at half-integral m. Another example is to create a polycrystal by assembling bits of gauze with random orientations on a glass plate.

Some more advanced topics which can be investigated with the same experimental apparatus are the following:

11. *Babinet's theorem* shows that the diffraction patterns of a mask and its complement are the same, except for a δ-function at the origin. Of course, experimentally this is not a δ-function, but the diffraction pattern of the unmasked illumination beam, which is approximately a Gaussian. Making the two masks is challenging. The experiment works best if both the mask and its complement are approximately 50% transmitting, and both must have some features which give strong diffraction detail outside the dimensions of the axial Gaussian spot. The masks used in the example shown in figure 4.6 were made as follows. First, the mask (a) was produced in thin metal sheet by drilling and filing. Then this was used as a contact mask in a vacuum evaporation system, through which aluminium was evaporated onto a piece of optically flat glass which then formed the mask (b).

12. *Annular aperture.* The importance of the annular aperture is that, of all real-positive masks, it has the smallest central spot for a given outside radius. The problem with an annular aperture is that the central obscuration has to be attached some way. Actually, if this is done with three thin wires, the result is indistinguishable from the ideal case. It is important that the width of the annular space be constant around the circle.

13. *Random arrays of holes*, or of identical apertures. To prepare an array which is really random, a random number table should be used to determine the positions of the holes, not just guesswork. Even so, some correlation will probably remain, because apertures cannot overlap.

14. *Quasicrystals.* Punch a set of holes at the vertices of a picture of the Penrose tiling pattern, which is based on two five-fold symmetric unit cells but has no translational symmetry [5]. The resulting ten-fold symmetric pattern observed in electron diffraction by quasi-crystals can be photographed, with spacing between the diffraction spots related to the golden ratio, $\tau = (1 + \sqrt{5})/2$. To get clear diffraction spots, one has to drill about 100 holes in the mask (figure 4.7). A variation on Penrose tiling is the Danzer quasi-periodic tiling,

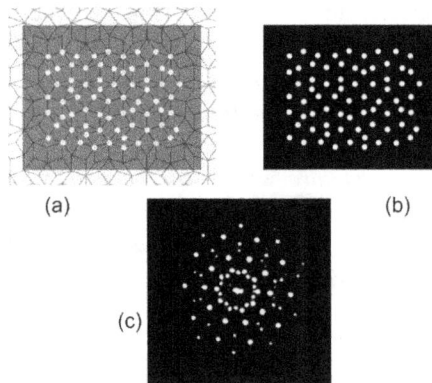

(a) (b) (c)

Figure 4.7. (a) Penrose tiling nodes, (b) the mask and (c) its diffraction pattern.

which has seven-fold symmetric tiles and gives a 14-fold symmetric diffraction pattern. A nine-fold tiling can also be found in the literature [6]. What is not obvious in these examples is why clear diffraction spots are obtained, despite the fact that the structures do not have translational symmetry. A possible answer is found by describing the quasiperiodic structure mathematically as an irrationally-oriented section of a periodic lattice in a higher dimension.

15. *DNA*. If a set of atoms lie at equidistant positions on a helix, the projection normal to the helix axis shows a set of points on a sine curve. A mask of this projection gives a diffraction pattern like the famous DNA x-ray diffraction pattern by Rosalind Franklin, which confirmed the double-helix model of Watson and Crick. How does the doubling of the helix affect the diffraction pattern (note that the translation between the two helices is not exactly half of the period)?

4.1.4 Comparison with calculated Fourier transforms

An important part of these experiments is the comparison between theory and experimental results. A computer code should be written in which an input mask picture is converted into a matrix, and then the two-dimensional Fourier transform is calculated and displayed. To get high enough detail in the transform, the mask matrix should be imbedded in a larger matrix of zeros, so that when taking the fast Fourier transform, the density of calculated points is sufficiently great. The observed diffraction pattern should be compared with the square modulus of the transform, although the absolute value often makes a better comparison because weaker details are emphasized.

4.2 Fresnel (near-field) diffraction

Historically, Fresnel diffraction preceded Fraunhofer diffraction, although today we can see that the latter is a special case which is easier to explain quantitatively. The fact that Fresnel's predictions (1819) could be justified experimentally was an important step in the establishment of the wave theory of light [7]. In Fresnel diffraction a coherent wave, not necessarily a plane wave, is incident on a mask and the continuing wave is observed on a screen or camera at an arbitrary distance from it (figure 4.8). No lenses are involved in the diffraction region after the mask.

From this, one immediately sees that Fraunhofer diffraction is the special case when the wave incident on the mask is a plane wave and the distance to the screen is infinite. In the laboratory, Fresnel diffraction is usually investigated by using a monochromatic point source to illuminate the object mask at distance z_1, and putting a receiving screen or camera chip at an additional distance z after the mask.

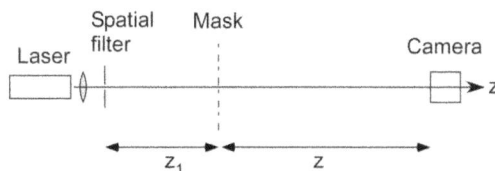

Figure 4.8. Setup for Fresnel diffraction.

The limit $z, z_1 \to \infty$ then corresponds to Fraunhofer diffraction, and $z \to 0$ to a geometrical shadow. Actually, the Fraunhofer condition can be more generally stated as $z^{-1} + z_1^{-1} = 0$, which means that $z = -z_1$ is a solution (see section 4.1.1). In general, a Fresnel diffraction pattern [2] of the mask $f(x,y)$ can be shown in the Fresnel approximation ($z \gg x, y$) to be the square modulus of the Fourier transform of the function $f(x, y)\exp[ik_0(x^2 + y^2)/2z_e]$, where $z^{-1} + z_1^{-1} = z_e^{-1}$, which makes simulation of the diffraction patterns quite straightforward. However, there are several special cases where the diffraction pattern or some feature of it can be calculated analytically, and some of these have scientific and historical importance. These are:

(a) Cases where the function $f(r)$ has axial symmetry, and can thus be expressed as a function of $r^2 = x^2 + y^2$. These include a circular disc, a circular hole and also a zone-plate, where $f(r)$ is a periodic function of r^2; in such cases, the diffraction pattern can be calculated analytically at points on the axis $r = 0$. Additionally, if $f(r)$ is a Gaussian function, the diffraction pattern can be calculated at all (x, y, z), not just on the axis, leading to an understanding of the propagation of a Gaussian beam.

(b) Cases where $f(r)$ is a simple function of one variable only, say x, and then in some cases the resulting integral can be evaluated. Examples are the diffraction pattern of a knife edge, where $f(x)$ is a step function, or a one-dimensional slit, where $f(x)$ is unity for a certain range of x and zero outside it. Here the integrals are not exactly analytic, but can be visualized by the use of an amplitude-phase diagram called the Cornu spiral (figure 4.9).

(c) Another example which is analytic is the Talbot effect (section 4.2.3), in which the diffraction pattern of a periodic one-dimensional object with period a can be shown to be identical to the mask function for certain values of z_e, which are multiples of the 'Talbot distance' $z_T = a^2/\lambda$.

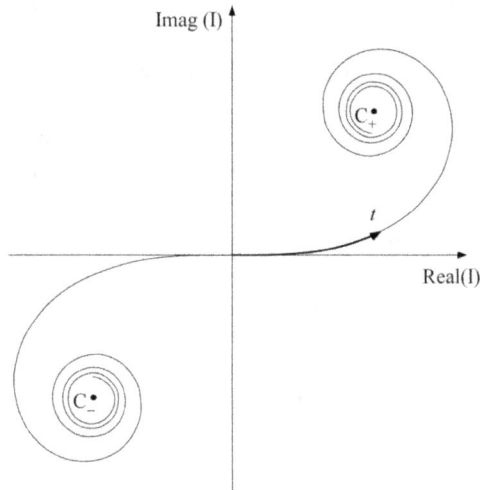

Figure 4.9. The Cornu spiral. The spiral is an amplitude-phase diagram representing the integral $I(s) = \int_0^s \exp(it^2/2)dt$ in which s and t are measured along the curve from the origin, and the electric field is proportional to Real $[I(s)] + i$ Imag $[I(s)]$. The points C_+ and C_- represent the values at $t = \infty$ and $t = -\infty$.

A convenient experimental system is straightforward (figure 4.8). A collimated spatially-filtered laser beam ($z_1 \to \infty$) falls on the object mask and the diffraction pattern at $z = z_e$ is projected onto a camera sensor (CCD chip) or onto a semi-transparent screen (tracing paper) which can then be photographed from the other side using a camera on axis. If the features on the mask lie within an inclusive circle of radius ρ, a good distance satisfies the relation that the variation in optical path between different points on the mask and the centre of the observation plane should be a few (say, five) wavelengths, i.e. the exponential factor $\exp[ik_0\rho^2/2z_e]$, should be about $\exp[10i\pi]$ at the edge of the mask, giving $z_e \approx \frac{\rho^2}{10\lambda}$. For example, if $\rho \sim 2$ mm, a suitable distance is about 1 m, and the expanded laser beam should be at least 8 mm in diameter so as to ensure uniform illumination over the mask area.

The experimental results should be compared to simulations in all cases. When carrying out simulations, it is important to realize that the function $f(x, y)$ can only be defined over a finite region of space. If $f(x, y)$ is not zero on the boundaries of this region, the simulation will assume that there is a boundary there, which produces its own diffraction pattern (see figure 4.11), which might obscure the effect being sought. It is therefore wise to multiply the incident field by a factor, such as a Gaussian, which falls to a small value at the boundary in order to avoid such artefacts.

4.2.1 Objects with axial symmetry

Several simple but important cases fall into this category, for which the Fourier integral $F(u, v)$ on the axis takes on the form $F(0, 0) = 2\pi \int_0^\infty f(r)\exp[i\frac{k_0}{2z}r^2]rdr$, which is often integrable analytically. Examples are round holes, Gaussian beams, annular apertures.

1. **The round disc.** A particularly important case is the round disc. In 1818, when Fresnel proposed the first theory of scalar-wave diffraction, it was at first rejected because Poisson pointed out that it would predict a bright spot at the centre of the shadow of a disc, which had never been observed. Arago and Fresnel performed the experiment carefully, and indeed observed the bright spot [7]. As a result of this, the theory came to be accepted. The experiment can best be carried out using a steel ball (2–3 mm diameter) as a mask, supported on a radial wire and blackened with soot to prevent reflections. At any distance, and for all wavelengths there is a bright spot at the centre of the diffraction pattern (figure 4.10), so the experiment should also be carried out using a white source (light emanating from the end of an optical fibre, for example). How accurately must the disc be round? When you have answered this question, you will understand why arbitrarily cut discs are unsuitable, and it is better to use steel balls, which are cheap but are manufactured to a high degree of accuracy. The corresponding simulations are shown in figure 4.11.

   Interestingly, the light reaching the central spot can be considered to have its origin in an 'edge wave' around the disc. By placing a camera in the diffraction pattern plane, with its aperture stop situated within the

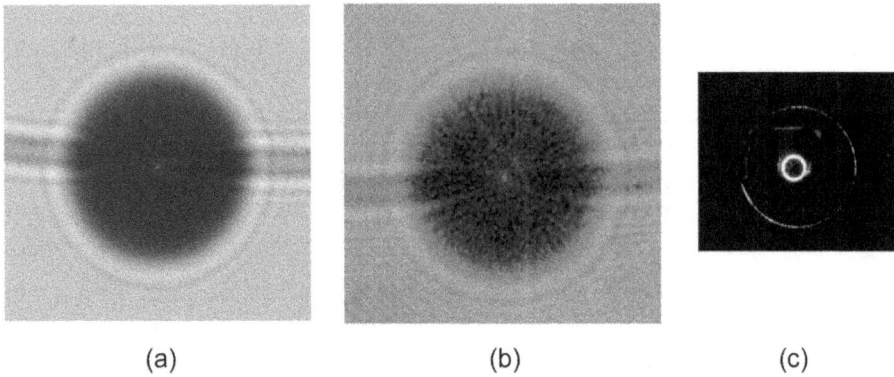

Figure 4.10. Fresnel diffraction by a 2 mm diameter steel ball, z_e = about 30 cm (a) with a white source, (b) with a laser source, (c) visualizing the edge wave.

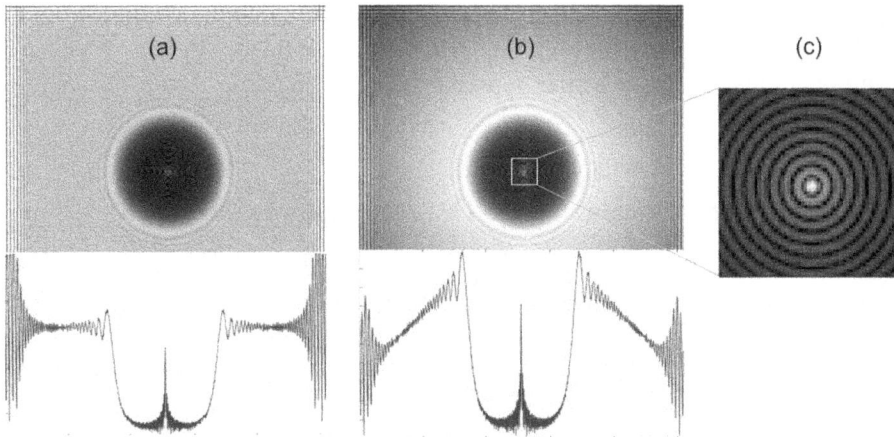

Figure 4.11. Simulations of the diffraction by an opaque disc with image profiles calculated through the central spot: (a) without modulation of the background and (b) with a Gaussian modulation. (c) Shows an enlargement of the central spot in (b).

geometrical shadow of the disc, and using its lens to focus on the ball, this edge wave can be photographed (figure 4.10(c)); the outer weak ring observed in the figure comes from light scattered by the aperture stop of the camera.

2. **Zone plate.** Another mask with axial symmetry, due originally to Fresnel, is the zone plate, which is a set of circular rings with radii proportional to the square roots of the natural numbers (figure 4.12(a)). Since the function is periodic in r^2, the mask creates point foci at specific distances (figure 4.12(b)). The transform depends on the variable $(\frac{1}{z} + \frac{1}{z_1}) \equiv \frac{1}{z_e}$, from which z_e behaves like a focal length, and so the zone plate is a 'diffractive lens'. This is the basis of the diffractive optics industry, which designs lenses with controllable parameters using Fresnel diffraction. Zone plates are also used as x-ray

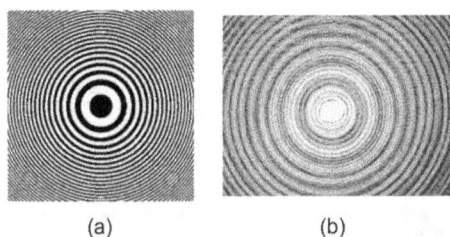

(a) (b)

Figure 4.12. (a) Zone plate and (b) one of its foci.

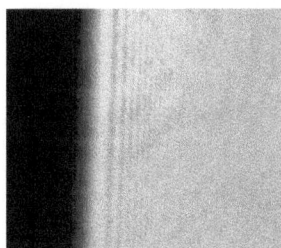

Figure 4.13. Diffraction by a straight edge.

lenses, since refractive lenses do not exist in the x-ray region, and reflective curved mirrors are hard to manufacture with sufficient accuracy [8]. The zone plate is also the physical concept behind holography, since it is a two-dimensional mask creating a point image located at a position in three-dimensional space; a hologram can be considered as the superposition of many zone-plates with different centres and values of z_1.

4.2.2 Linear objects: knife edge and slits

The integrals describing diffraction by linear objects can rarely be evaluated analytically, so the experimental results must be compared with simulations. A basic example is diffraction by a knife-edge, or step-function $f(x, y) = 1(x > 0)$, $f(x) = 0$ $(x < 0)$, shown in figure 4.13. Various slits can also be investigated. In this case, the edge wave construction can be used to interpret the pattern approximately in the bright region; it is the interference between a cylindrical wave (scattered by the edge) and a plane wave. In the dark region it is the cylindrical wave alone. But the Cornu spiral method (figure 4.9) shows the origin of the diffraction pattern most clearly; the intensity measured as a function of x in figure 4.13 is the square of the vector from point C_- to t when t takes values of x going from negative to positive.

4.2.3 Fresnel diffraction by a one-dimensional periodic object: Talbot re-imaging effect

An interesting effect which was discovered by Talbot [9] in 1836 relates to the Fresnel diffraction pattern of a periodic linear mask or diffraction grating (figure 4.14). For the case $z_1 \to \infty$, we consider a plane wave incident on the

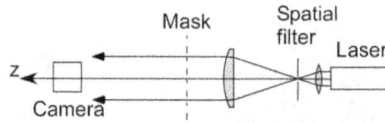

Figure 4.14. Set-up for Talbot re-imaging.

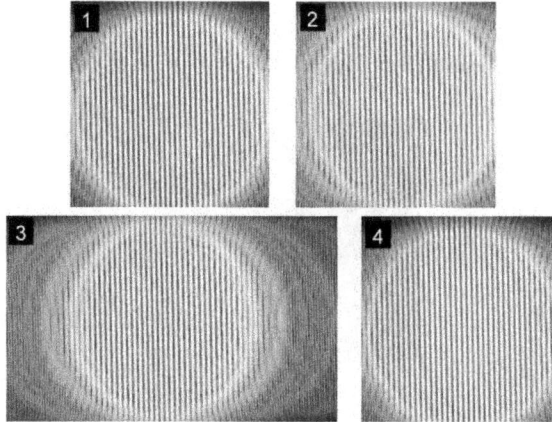

Figure 4.15. The first four Talbot images of a 5 line/mm Ronchi grating. The third image shows the overlapping orders of diffraction.

mask, which is basically a diffraction grating of period a, so that the diffracted wave can be considered as a discrete array of plane-wave diffraction orders. These waves superimpose in a subsequent plane to give the Fresnel diffraction pattern. The diffracted wave of order m propagates at discrete angle θ_m, where in the paraxial Fresnel approximation $\theta_m \approx m\lambda/a$, and so after propagation to a plane at distance z from the grating its phase is advanced by $\delta\phi = \vec{k}\cdot\vec{z} = k_0 z \cos\theta_m \approx k_0 z(1 - \frac{1}{2}\theta_m^2)$. At certain distances the phase-shifts $k_0 z \theta_m^2/2$ of the superimposed waves are all integer multiples of 2π, so that the Fresnel diffraction patterns are exact replicas of the original grating. These distances are *even* multiples of the 'Talbot distance' defined as $z_T = a^2/\lambda$. At *odd* multiples of z_T the diffraction patterns are exact replicas shifted by half a period, since successive diffraction orders are phase-shifted by alternately even and odd multiples of π. For example, using light with $\lambda = 633$ nm, a grating with $a = 0.2$ mm has $z_T = 63$ mm. The Talbot images have the interesting property that non-periodic errors are not reimaged, and there are some applications of this idea in photolithography. Note that two-dimensional period objects are only reimaged in specific planes when their periods in the x and y directions are rationally related.

The pictures shown in figure 4.15 show successive images of a 0.2 mm-period Ronchi grating using a He–Ne laser, for which the Talbot distance is 63 mm. How can we demonstrate the lateral shift of the 2nd and 4th images by half a period relative to the 1st and 3rd? See section 4.2.4.

Simulations of the Talbot effect usually show a plot of the x-profile of a one-dimensional mask and its reconstructions as a function of the propagation distance, along the z-axis. The picture repeats after a propagation distance of $z = 2z_T$. It is known as the 'Talbot carpet'. This is difficult to observe directly in optics, but is a computation representing the expected diffraction patterns. It is easy to see the shifted reproductions after propagation distances of odd multiples of z_T. (figure 4.16). The Talbot carpet has recently been experimentally visualized in water waves by Bakman *et al* [10].

The question arises: how did Talbot see this effect in 1836? If you read Talbot's original paper, you will see that he used a distant white point source of light to illuminate the grating, and observed the transmitted light field with a magnifying glass at different distances. He saw that the grating image was reconstructed in different *colours* at different distances, and the process repeated itself periodically. It is very interesting to repeat the experiments described in the paper; remember that the explanation in terms of wave theory was only provided by Rayleigh [11] in 1881, 45 years later.

4.2.4 Radial star target

The Fresnel diffraction pattern of a radial target (figure 4.17), which appears quite complicated, can be understood quite well in principle in terms of Talbot re-imaging.

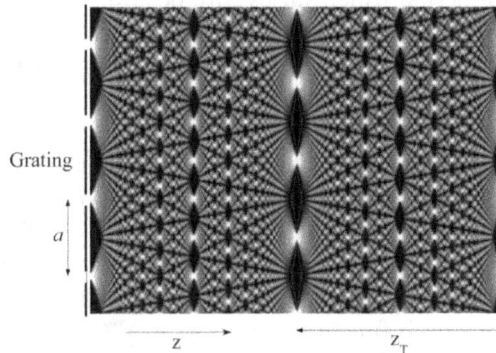

Figure 4.16. A simulation of the Talbot carpet.

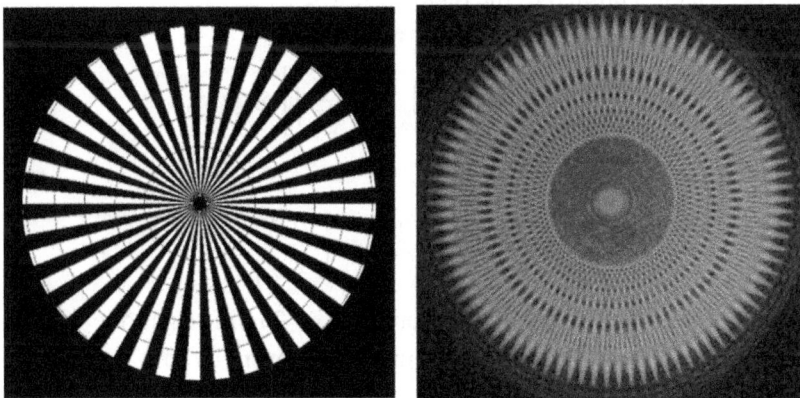

Figure 4.17. A radial target with 10° period and an example of its Fresnel diffraction pattern.

Remember that the details of the pattern are a function of z_e, where $z^{-1} + z_1^{-1} = z_e^{-1}$. It has a close relationship to the Talbot carpet, as you can see by comparing figure 4.17 to figure 4.16. In this case, the period a is not a constant, but at radius r has a value $a = 2\pi r/N$, where N is the number of spokes in the star target, so that at given λ the Talbot distance is a function of the radius $z_T = a^2/\lambda = 4\pi^2 r^2/N\lambda$. The repeat images occur at equal intervals of z_T, i.e. $z_e = mz_T$, where m is an integer. For a given z_e it follows that they occur at radii given by $r^2 = z_e N\lambda/4\pi^2 m$. Repeat images at radii proportional to $1/\sqrt{m}$ are evident in the diffraction pattern in figure 4.17, and you can also easily see the shift of half a period between even and odd values of m.

4.3 Diffraction gratings: transmission and reflection gratings and spectroscopy

The diffraction grating is an optical element which has transmittance or reflectance which is a periodic function of position. In its simplest form, it can be considered as a set of uniformly-spaced slits, with period d in one dimension, although more complicated and useful versions may have periodicity in more than one dimension, the 'slit' might have a complex (phase-changing) transmission or reflectance, the structure might have a period which varies with position, or the grating might be on a spherical or cylindrical surface. The basic mathematical description of a grating is a transmittance or reflectance function which is a set of delta-functions representing the positions of the slits, convolved with a second function which describes the properties of the individual slit.

The diffraction by the grating is observed in the Fraunhofer regime, using a collimated beam of incident light. The diffracted wave is observed in the focal plane of a converging lens or lens combination (figure 4.18). The light distribution observed is therefore the Fourier transform of the incident light wave multiplied by the grating function. For a finite grating in one dimension, the set of delta-functions is an infinite comb function, $\mathrm{comb}(\frac{x}{d}) \equiv \sum_{n=-\infty}^{\infty} \delta(x - nd)$, multiplied by an aperture of extent L representing the finite length of the grating, $\mathrm{rect}(\frac{x}{L})$. The grating function is then a convolution between this finite comb and the individual slit function $g(x)$ which is limited to the region $0 < x < d$, having zero value outside this range. The grating function is therefore

$$f(x) = \left[\mathrm{comb}\left(\frac{x}{d}\right) \cdot \mathrm{rect}\left(\frac{x}{L}\right) \right] \otimes g(x),$$

whose Fourier transform is

$$F(u) = \left[\mathrm{comb}\left(\frac{ud}{2\pi}\right) \otimes \mathrm{sinc}\left(\frac{uL}{2}\right) \right] \cdot G(u).$$

This represents the set of diffraction orders at $u_m = 2\pi m/d$, each one broadened by the function $\mathrm{sinc}(\frac{uL}{2})$ and with peak amplitude given by the transform of the individual slit, $G(u_m)$ (figure 4.19(a)). Note that the sequential order in which the convolution and multiplication operators are carried out, indicated by the positions

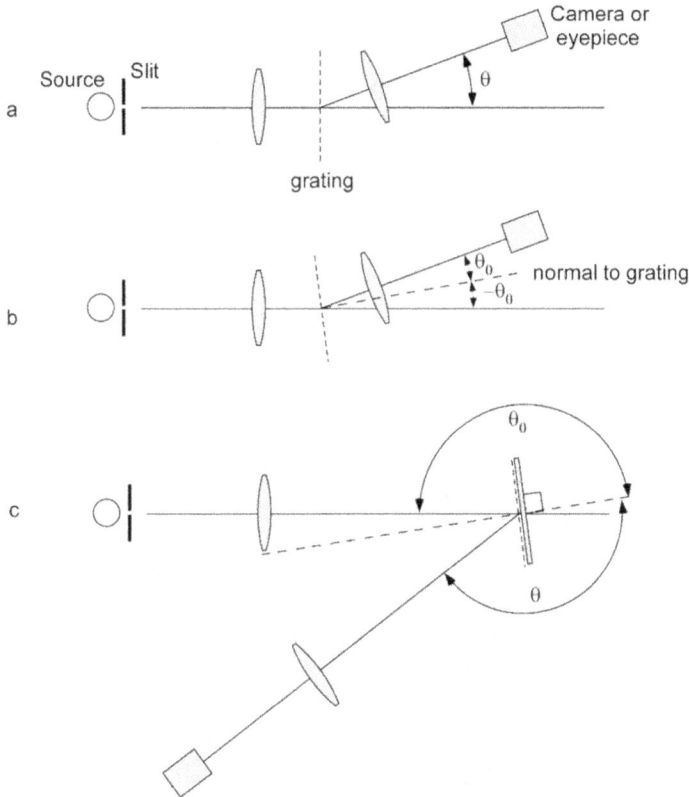

Figure 4.18. Three standard observation and measurement regimes: (a) normal incidence on a transmission grating, (b) minimum deviation by a transmission grating, (c) using a reflection grating.

of the square brackets in the above expressions, is important. The sinc function indicates the width of the orders, since it has basic half-width (value of u at the first zeros near the origin) $\Delta u = \frac{2\pi}{L} = \frac{2\pi}{Nd}$, where N is the number of slits exposed to the incident beam. This defines the resolution achievable by the grating to be $\frac{\Delta u}{u_m} = \frac{2\pi}{Nd}\frac{d}{2\pi m} = \frac{1}{Nm}$. The 'resolving power' of the grating is the inverse of this ('larger is better'):

$$R \equiv \frac{u}{\Delta u} = Nm.$$

This is a basic formula which applies to any interferometric instrument: the *resolving power is the product of the number of interfering waves and the order of interference*, the order being defined as the path difference between the individual interfering waves, in units of the wavelength.

4.3.1 Square wave grating

This is an interesting example of a slit grating, a 'Ronchi ruling', in which the slit widths are exactly half of the spacing between them. This grating has zero order and

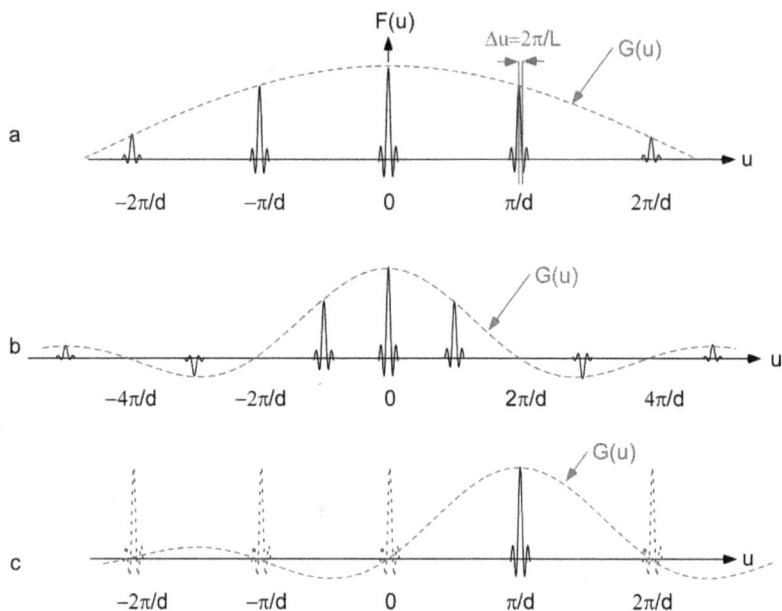

Figure 4.19. (a) The amplitude spectrum of a grating with arbitrary real $g(x)$, illustrating the resolution; (b) The example of a square-wave grating (Ronchi ruling) where the even orders have zero amplitude; (c) The example of a blazed grating at the design wavelength, for which only the first order is non-zero.

odd orders, but no even orders (figure 4.19(b)). It is possible to vary the slit width by superimposing two such gratings parallel to one another with variable shift.

4.3.2 Blazed gratings

Blazed gratings are gratings in which $g(x)$ is designed to concentrate the diffracted energy into a specific order, usually $m = 1$. Since this order has $u_1 = 2\pi/d$, we require $G(\frac{2\pi m}{d}) = 0$ for all m except $m = 1$.

This diffraction pattern is not symmetrical about $u = 0$, therefore $g(x)$ must be a complex function. The requirement is satisfied by $G(u) = \text{sinc}(\frac{ud}{2} - \pi)$, (figure 4.19(c)), which is the transform of $g(x) = \text{rect}(\frac{x}{d}) \cdot \exp(\frac{i2\pi x}{d})$, so that the grating function has constant amplitude, but a phase which has the form of a saw-tooth wave changing linearly by 2π in every period (figure 4.20). Physically, each 'prism' of this grating deviates the incident wave by exactly the angle of the first order. This can be true only for a given wavelength, although it will be approximately true for surrounding wavelengths too. For this reason a commercial blazed grating always indicates the design wavelength, at which the diffraction efficiency is close to 100%.

4.3.3 Spectroscopy

Both transmission and reflection gratings (figure 4.18) are widely used for spectroscopy. The former gratings are most common in teaching laboratories, but the latter are ubiquitous in serious spectrometers. In general, we have the diffraction grating

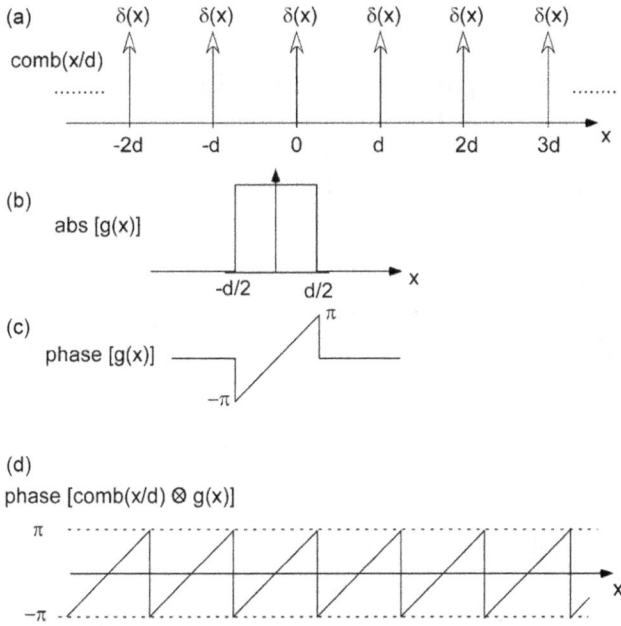

Figure 4.20. The basic description of a blazed phase grating. (a) the periodic array of δ-functions, with period d, (b) the transmittance or reflectance within one period, (c) the phase within one period, (d) the phase as a function of position in the complete grating.

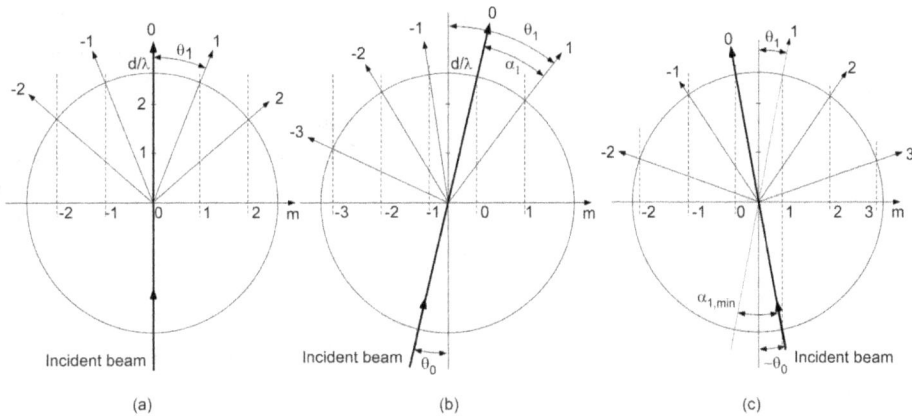

Figure 4.21. (a) Geometrical construction for grating diffraction, at normal incidence (from below); (b) oblique incidence at θ_0; (c) minimum deviation in the first order.

equation for incidence at angle θ_0 to the grating normal, and diffraction into the mth order at angle θ_m (figure 4.21):

$$m\lambda = d(\sin\theta_m - \sin\theta_0) \tag{4.1}$$

where d is the grating period and λ is the wavelength. This can also be written in terms of the deviation $\alpha_m = \theta_m - \theta_0$ as $m\lambda = d(\sin(\alpha_m + \theta_0) - \sin\theta_0)$.

A transmission grating gives minimum deviation $\alpha_{min} = 2\theta_0$ when $\theta_0 = -\theta_m$ (figure 4.18(b) and 4.21(c)).

A reflection grating has θ_0 around $180°$ (figure 4.18(c)). It has several applications where $\theta_0 = \pi + \theta_m$, under which conditions the mth order reflection returns along the direction of incidence (for example, in tunable wavelength laser cavities). Moreover, most reflection gratings are blazed, to achieve high diffraction efficiency in a given (usually first) order. They do not show minimum deviation. As a result, the angle of incidence can in principle be used to tune the grating to have maximum efficiency for a chosen wavelength.

The theoretical resolution achievable with a grating depends on its length L, or the length which is illuminated. As shown above, the resolving power achievable in any interferometric system is given by $R = mN$, where m is the order of diffraction and N the number of interfering waves. The interpretation of R depends on the purpose of the interferometer; if it is a grating used for spectroscopy $R = \lambda/\delta\lambda$ where $\delta\lambda$ is the smallest wavelength difference resolvable, or if an interferometer is used to measure changes in length x of an object, $R = x/\delta x$. If a diffraction grating is used in the maximum order possible, from equation (4.1) we deduce that $R = m_{max}N = 2L/\lambda$ independent of the number of slits, which is surprising. Young's slits, with a separation L, give the same resolution as a diffraction grating with that length! The differences, of course, lie in the light throughput and convenience of interpreting the spectrum.

Setting up the spectrometer for a transmission grating is not difficult. First, the telescope eyepiece should be focused so that the user sees the cross-hairs sharply (this adjustment depends on the observer's eye). Then, with no grating, the telescope should be turned to a direction which allows it to focus on a distant object, and its length should be adjusted to give a sharp image[2]. Following this, the telescope is turned to observe the spectrometer slit and the collimator length is adjusted to give a sharp image of the slit centred in the field of view. The slit should now be aligned to be vertical (by eye), the light source placed as close as practical to the slit, and its position arranged to give maximum intensity as observed through the telescope. The slit width can then be reduced to its minimum. Now the grating should be put in place. To ensure that its grooves are vertical, the grating should be rotated in its own plane till the images of the slit in all orders of diffraction appear at the same height. The spectrometer probably has levelling screws which will allow this adjustment to be made in a controlled way, using the top or bottom of the image of the slit, or some irregularity in the slit jaws, to judge the relative heights of the images. All these procedures can all be carried out using a camera in place of the eyepiece.

The following experiments can be done using a Hg, Na or other discharge lamp, and a transmission grating which is not blazed and shows all orders. 600 lines per mm is a typical grating period and this number is used in the examples below.

1. Using a singlet line (e.g. 540 nm from Hg) measure the *minimum deviation* α_m in each accessible order ($-4 \leqslant m \leqslant 4$) and confirm that $m\lambda = 2d\,\sin(\alpha_m/2)$.

---

[2] The 'object at infinity' prepared for the experiments on geometrical optics in sections 2.3 and 2.4 can also be used for this purpose.

2. For normal incidence, confirm that $m\lambda = d \sin \theta_m$. It is problematic to achieve $\theta_0 = 0$ exactly, but this is not important since, from equation (4.1), if you have a constant error θ_0 and plot $\sin \theta_m$ versus m you should still get a linear plot which goes through the point $(0, \sin \theta_0)$.
3. Use a Ronchi Ruling as a grating, and observe the many orders of diffraction. Note that the even orders (except for zero) are missing.
4. The Na doublet 589.0 and 589.6 nm requires $R = 0.6/589 \approx 1000$ to resolve the components. This can be confirmed by introducing a variable slit before the grating. For example, if this variable slit has width 1 mm and then $N = 600$, $mN = 600m$ and the doublet should not be resolved in the first order, but should be resolved in the second and higher orders.
5. After calibrating the grating using, for example, the known wavelengths of the Hg spectral lines, the wavelength of an 'unknown' source can be measured. For example, this can be a He–Ne laser, the value of whose wavelength is important in later experiments.
6. The same system is used in experiments on laser modes (section 10.3).

Using a blazed reflection grating, the intensities of the various orders can be measured and the data used to fit a model for the grating facet angle, which determines the envelope function $G(u)$, assuming that the phase profile (figure 4.20(c)) is a simple saw-tooth function, but the wavelength used may not be the design wavelength. An example of the first orders from a blazed grating with a green LED source are shown in figure 4.22.

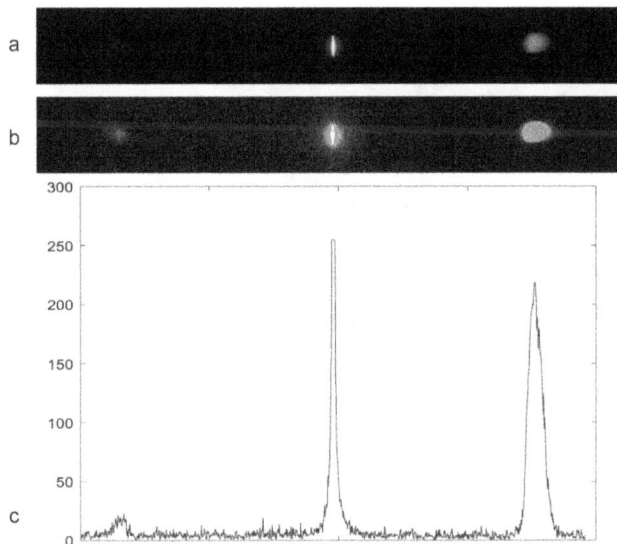

Figure 4.22. Spectrum from a green LED using a reflection grating blazed for the first order. (a) recorded image of the −1, 0 and 1st orders, (b) contrast-enhanced image, (c) profile of (a) through the three orders.

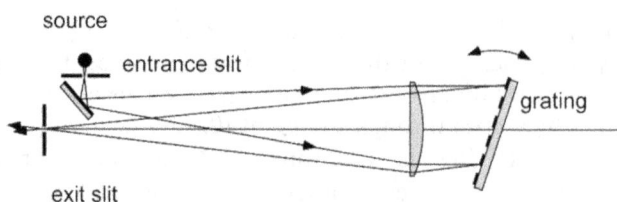

Figure 4.23. Construction of a monochromator.

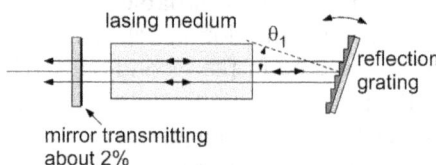

Figure 4.24. Construction of a tunable diode laser.

4.3.4 Monochromator

A reflection grating can be used as a monochromator, i.e. an instrument which separates a given wavelength from a source with a broad spectrum, such as a white lamp. The lamp-light is focused onto a slit, which is placed at the focal point of a converging lens so as to produce a collimated beam on the grating. The same lens or a different one is used to collect the diffracted light and return it to the output slit (figure 4.23). Rotating the grating scans the wavelength scale. The bandwidth of the selected light can be chosen by changing the slit width. The monochromator can also be used in place of a very high resolution variable filter in experiments where there is space for it.

This is also the mechanism used in tunable laser diodes; the grating replaces one of the mirrors of an external optical resonator whose mode frequencies decide the frequency of oscillation of the diode laser (figure 4.24); rotating the grating, usually by a piezo-electric device, changes the output frequency of the laser within the bandwidth of the lasing material. It is important, to achieve a high-gain resonator, that the blazed reflection grating should be optimized for the typical output wavelength region.

4.4 Imaging with coherent illumination

Almost all everyday optical instruments are used for imaging. The process of producing an image is well understood from the point of view of geometrical optics, but its limitations eventually result from the wave nature of light. In particular, the limit of resolution—the smallest distance between two details which can be discerned as separate using a given optical system—has to be understood using physical optics. Ever since Ernst Abbe's pioneering ideas, formulated in 1873, in which he suggested that the limiting resolution of a microscope was determined by the wavelength of the light used and the angular aperture of the objective lens, the subject of microscope

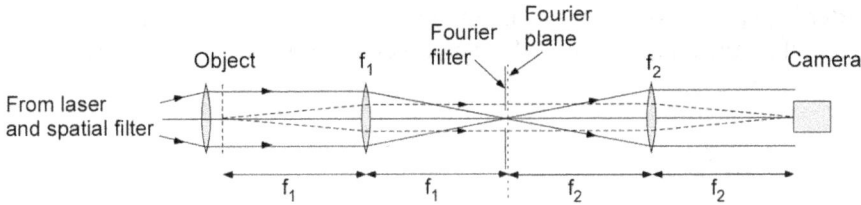

Figure 4.25. 4-f imaging system with magnification $-f_2/f_1$.

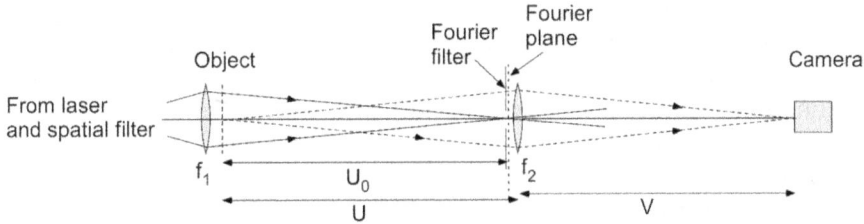

Figure 4.26. Abbe–Porter imaging system with magnification $-V/U$.

and telescope imaging has been a central topic in physical optics. Recent years have seen substantial improvements in resolution which have improved the resolution limit by more than an order of magnitude using techniques for which the Nobel prize was awarded in 2014.

4.4.1 Coherent imaging experimental setups

Although most imaging is carried out using incoherent light (as, for example, in conventional microscopes or telescopes), the limitations of imaging systems and some of the methods for image processing can best be understood when using coherent illumination. The principles can afterwards be extended to the incoherent case. We therefore consider the image of a light-transmitting slide whose transmittance is described mathematically by $f(\vec{x})$, where $\vec{x} \equiv (x, y)$, illuminated by means of a coherent plane wave from a laser source. The imaging process is described theoretically by a double Fourier transform. $F(\vec{k})$, the Fourier transform of $f(\vec{x})$, is first produced in an intermediate plane of the imaging system, and then the continued propagation of the light waves creates a Fourier transform of $F(\vec{k})$ which is $f(-\vec{x})$. Here, a system with (minus) unit magnification is assumed. The idea is most easily described for a '4-f system' (figure 4.25), in which the two Fourier transforms are carried out using lenses with different focal lengths, so that some magnification is obtained. However, from the experimental point of view, the 4-f system is not the easiest to use, because both lenses f_1 and f_2 are involved in the imaging stage, and an earlier version, the 'Abbe–Porter' system (figure 4.26), used by Porter [12] in 1906 to demonstrate Abbe's theory of image formation, separates the two Fourier transformations and is easier to set up and adjust.

In the 4-f system, the scale of the spatial frequencies in the Fourier transform plane is proportional to λf_1, and in the Abbe–Porter system to λU_0. The advantage of the latter is that the Fourier plane magnification can be adjusted and aligned before insertion of f_2, the imaging lens, which can then be chosen independently to give the required magnification.

4.4.2 Resolution limit

According to the Abbe theory, the resolution of the system is determined by the highest spatial frequency transmitted by the aperture stop. This stop is an iris diaphragm placed in the Fourier plane, whose diameter can be varied continuously. The change in resolution with the stop diameter is vividly demonstrated in figure 4.27 using a radial target (figure 4.17) as object, for which the local spatial frequency is inversely proportional to the radial distance from the centre. This radial target is more intuitively understood than the standard USAF resolution target (see figure 4.35). It is always a surprise to see that as one closes the Fourier filter iris, the circular 'blur circle' in the centre of the image, where there is no resolution, expands! See the video clip 'closing the iris' in figure 4.27.

Quantitatively, this experiment will confirm that the *coherent* resolution limit is λ/NA, where NA is the numerical aperture. The NA is defined as the sine of the semi-angle subtended by the aperture stop of the imaging lens at the object plane.

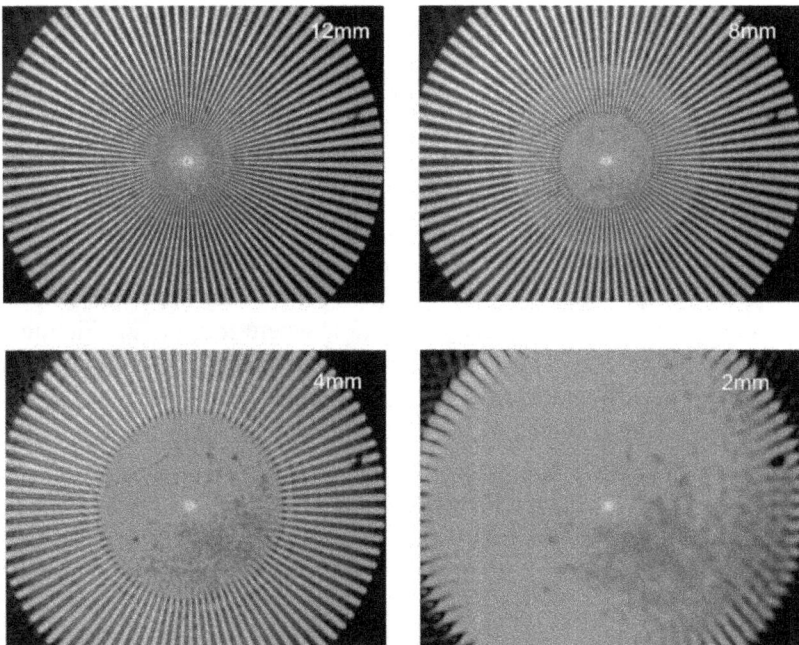

Figure 4.27. Coherent images of the radial target with various iris radii (indicated on each image) with a distance $U_0 = 500$ mm between the object and the Fourier plane. The outer radius of the target is 5 mm. The influence of the iris diameter on the resolution limit is clear. In the 8 mm image you can discern a change in the images at twice the resolution limit. Where does this come from? Video available online at http://iopscience. iop.org/book/978-0-7503-2300-0.

4.4.3 Passive resolution improvement

The system described above can be used to illustrate ways to improve the resolution. We call this 'passive' because it does not involve any physical modifications to the object, such as the attachment of fluorescent molecules, which are the basis of the super-resolution techniques (STORM and STED) which were honoured by Nobel prizes in 2014.

1. The idea implemented by Abbe was to observe the image while illuminating the object incoherently from a range of angles. The results of the coherent imaging (above) for a given aperture size can be compared in the Abbe–Porter system with the image obtained with the laser/spatial-filter replaced by a LED or flashlight with its beam expanded to illuminate the whole target from a range of angles in a similar way. Abbe theory shows that improvement of the resolution by a factor of 2 can be obtained this way. This idea is illustrated in section 4.5.

2. A recent advance in imaging is 'Ptychographic Microscopy' (section 4.7), in which a high resolution two-dimensional image is synthesized by combining several low-resolution images with different illumination directions. These illumination directions correspond to different off-centre small apertures in the Fourier plane. The high resolution image can be obtained by using each low resolution image to create a partial diffraction pattern, and these patterns are stitched together to create a larger pattern which improves the resolution of the image. To stitch them together, the phases have to be known, and these are found by using phase retrieval (section 4.5.3). The technique is described further in section 4.7.

4.4.4 Spatial Filtering in the Fourier plane[3]

The spatial filter is a mask which is inserted into the Fourier plane and has the effect of modifying the function $F(\vec{k})$ before it is retransformed to provide the final image. For example, in the resolution experiments shown above, the spatial filter is the iris diaphragm centred on the optical axis, which transmits light in that plane up to a certain value of $|k|$ and blocks it at larger values. Mathematically, the filter is described in general by a complex function $G(\vec{k})$ (its complex nature allows phase changes to be implemented), so that the image obtained is the Fourier transform of $F(\vec{k}) \cdot G(\vec{k})$. This transform is the convolution $f(-x) \otimes g(-x)$, although in many cases this is not the easiest way to envisage it. Porter's experiments [12] (section 4.4.5) illustrated the principle by imaging a piece of gauze and then blocking various orders of the diffraction pattern in the Fourier plane, which resulted in different images and artefacts which can easily be understood as reconstructions from the

---

[3] The concept of 'Spatial Filter' here must be distinguished from the spatial filter used to clean up and expand the laser beam before it illuminates the object, as in section 1.2.2 and figures 4.25 and 4.26, although the principle is the same.

orders which were transmitted. This basic setup can be used to illustrate various aspects of coherent imaging. Some suggested experiments are:

1. Removal of periodic artefacts from an image, such as the printing screen used in newspaper photographs.
2. If the object slide shows a monkey behind periodic bars in a cage, it can be 'released' by filtering out the Fourier transform of the bars.
3. Dark ground imaging, in which a weak object on a strong uniform background can be emphasized.
4. Schlieren imaging, to emphasize phase gradients.
5. Phase imaging by the Zernike technique, widely used in biological imaging.
6. Shadowgraph imaging.

1–2. Sometimes a specific spatial frequency must be removed from an image. For example, if you look at an image printed in a book or newspaper through a magnifier you will see that it is actually a lattice of dots of varying sizes in accordance with the intensity of the image at each point (called a 'half-tone' image). When this lattice is periodic, it can be filtered out from the picture by an appropriate filter in the Fourier plane in order to give a continuously varying image, which has some aesthetic advantages although it does not improve the resolution. Another example is 'removing the monkey from its cage', by blocking the transform of the periodic bars (figure 4.28, parts 1 and 2).

3. **Dark ground imaging** is useful when a weak object is superimposed on a bright background field containing little information of interest. It was originally used by Brown in observing Brownian motion of small particles moving in a fluid as the result of thermal excitation, and he called it an 'ultramicroscope'. We demonstrate this experimentally in the system of figure 4.26 by using as object a transparent plate with some small black marks on it, such as a reticle scale (figure 4.29(a)). The Fourier plane is then dominated by a strong peak on the axis, formed by all the light which is transmitted between the marks. This can be eliminated by a mask which should be a small circular black dot on a clean optically flat window (figure 4.28, part 3), adjusted in position to block the peak. The adjustment in all three dimensions is critical. When the axial position is not correct, the dark ground is obtained only in a central region of the image; this region expands as the correct position is approached. If the object marks are small, they contribute uniformly to the whole Fourier transform, and therefore most of their diffracted light bypasses the mask spot and creates an image of bright marks at the positions where the original black marks were (a negative image: figure 4.29(b)). If the black marks are not small in area compared to the transmitting region, the result is a bit more complicated.

4. **Schlieren imaging** system. This form of imaging emphasizes phase gradients in a transparent object. It is well-known in wind-tunnel optics, where it is used to visualize air-flow, by using the dependence of refractive index on pressure. Here, we use the system of figure 4.26 and insert a knife-edge in the Fourier plane (figure 4.28, part 4). First, the knife-edge is adjusted without an object so as to just obstruct the focal spot, creating a dark ground. This requires critical adjustment of the knife

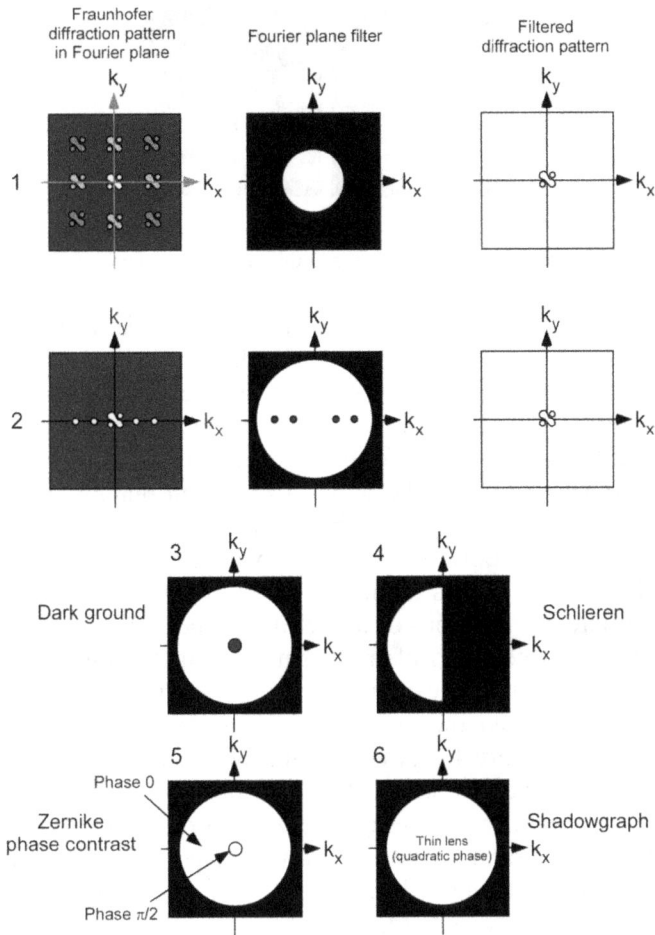

Figure 4.28. Fourier plane filters used in the above experiments.

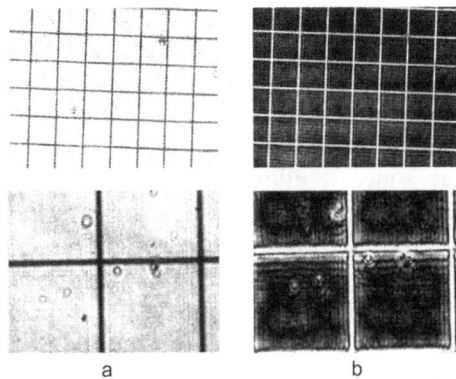

Figure 4.29. Dark field imaging. Images of lines in a 0.5 mm period reticle scale at two different magnifications, obtained with the system of figure 4.26 using two different imaging lenses: (a) bright-field and (b) dark-field images of the same regions.

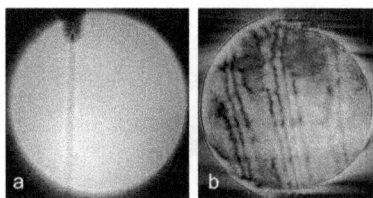

Figure 4.30. Schlieren images (a) of petroleum vapour squeezed out of a wash-bottle and (b) of a standard glass microscope slide in the object plane. The ripples in thickness are a common feature of glass plates.

position, and is a very sensitive test of the lens quality and cleanliness! In fact, the same system, known as the 'Foucault knife-edge test', is used for lens-testing. Suitable phase objects are a fingerprint or some oil or glue wiped onto a slide, or the gas ejected by a propane lighter (unsparked!) or by an airo-duster canister, since the gas has a different refractive index from the surrounding air, or simply the hot air flow around a soldering iron. If the optics are of high quality and well adjusted, images of aerodynamical effects can be obtained with air blowing across an obstruction, since the air cools because of its expansion, and the cooling increases the refractive index. The images obtained with different orientations of the knife-edge illustrate different projections of the refractive index gradient vector. The idea is easy to visualize using geometrical optics, since a refractive index gradient acts like a thin prism, but an analytical explanation using Fourier transforms is not much more complicated and then the image can be understood mathematically. Two examples are shown in figure 4.30, made with a system which used standard lenses, 25 mm diameter and a vertical knife-edge in the focal plane. Figure 4.30(a) shows visualization of petroleum vapour being squeezed out of the nozzle of a wash-bottle. Figure 4.30(b) shows the same field with a standard glass microscope slide inserted in the sample plane, where the ripples in thickness, typical of a glass slide which has not been optically polished, are visualized. The Schlieren system emphasizes the gradients in optical thickness in the x-direction.

5. **Zernike phase contrast**. Many objects such as biological cells are transparent, but they are assembled from materials with different refractive indices. As a result, the waves transmitted or reflected by them contain information in their phase, rather than intensity. A phase-contrast microscope, invented by Zernike in 1935, converts phase information into intensity information by modifying the phase of the zero order of the Fourier transform by about $\pi/2$. In 1953, Zernike received the Nobel Prize for this technique, since it allowed for the first time observation of live biological samples under a microscope without staining. A Fourier-plane mask to do this approximately can be made from a small drop of dilute lacquer allowed to dry on a glass slide, or by making a small hole in a thin plastic sheet (figure 4.28, part 5); a high-quality filter can only be made by controlled evaporation of a transparent film (e.g. MgF_2) on glass. A suitable phase object could be made by smearing some oil or transparent glue on a slide, or just using a piece of plastic sheet which has been extruded so that its thickness is not uniform. An important feature of the Zernike method is that the observed intensity is linearly related to the phase for small values,

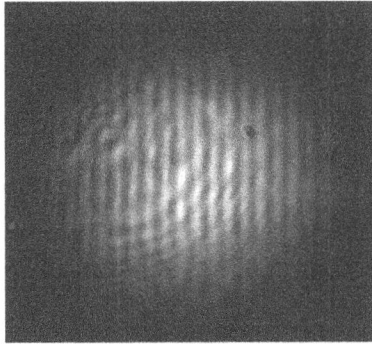

Figure 4.31. Shadowgraph image of a resonant acoustic wave in water. The regions which are bright and dark have lower and higher pressure than the average.

and therefore the method can be used to get quantitative high-resolution phase images in this approximation. Dark field imaging can also show phase differences, but in that case the intensity change in the image is proportional to the square of the phase for small values.

6. **Shadowgraph imaging**. This method, which is widely used for visualizing dynamic phase objects, uses a small defocusing of the imaging system. In Fourier space this is equivalent to multiplying the transform by a phase factor $\exp(i\alpha |k|^2)$ (figure 4.28, part 6). The constant α depends on the degree of defocus, which also degrades the resolution of the imaging system so that, although it is easy to implement, the method is not suitable for high resolution phase imaging. Figure 4.31 shows such an image of an acoustic pressure wave in a fluid (see section 6.2).

4.4.5 Demonstrating spatial filtering

Porter's setup (figure 4.26) can easily be used for a dynamic demonstration of coherent imaging, in which both the spatially-filtered Fourier transform and the resultant image are projected side-by-side. This is very useful in developing physical insight into the Fourier transform process. It is done by inserting a beam-splitter between the Fourier plane and the imaging lens. One output from the beam-splitter is used to create the image of the object and the other output to image the Fourier plane, as shown in figure 4.32. A suitable object for this demonstration is a piece of gauze. This is roughly periodic, but contains errors, so that the Fourier transform is a square array of points which are not δ-functions. Using slits, circular apertures, opaque strips and other filters in the Fourier plane, the image can be changed significantly (figure 4.33). In particular, the non-periodic errors (blocked holes, for example) can be imaged using a filter which isolates a single order in the array (d, i). Also, by isolating the five central points of the transform with a circular hole filter (e), artefacts in the image can be created (j). Such artefacts have sometimes been interpreted erroneously as real image information in electron micrographs of periodic objects whose period is close to the resolution limit!

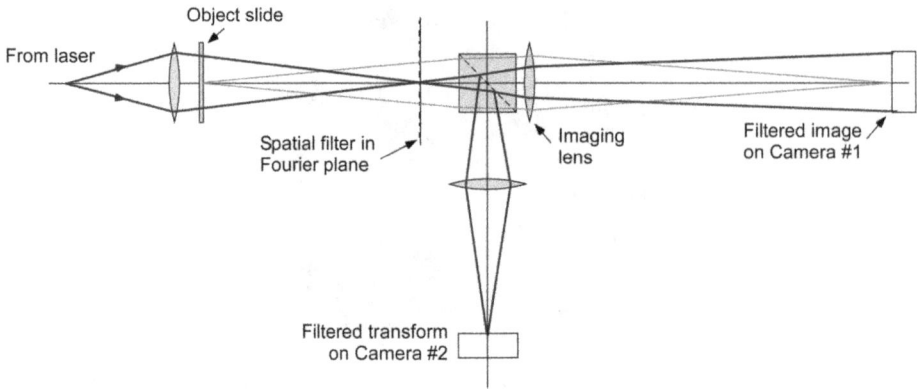

Figure 4.32. Setup for demonstrating spatial filtering in coherent imaging. If the two separate camera images cannot be displayed simultaneously, a plane mirror can easily be added in one of the two exiting beams so as to project the two images side by side on the same camera. Various filters can be used (such as 1–6 above), and topics such as sensitivity to the orientation of the filter or its correct alignment, can easily be demonstrated (reference [2], figures 12.5 and B.5).

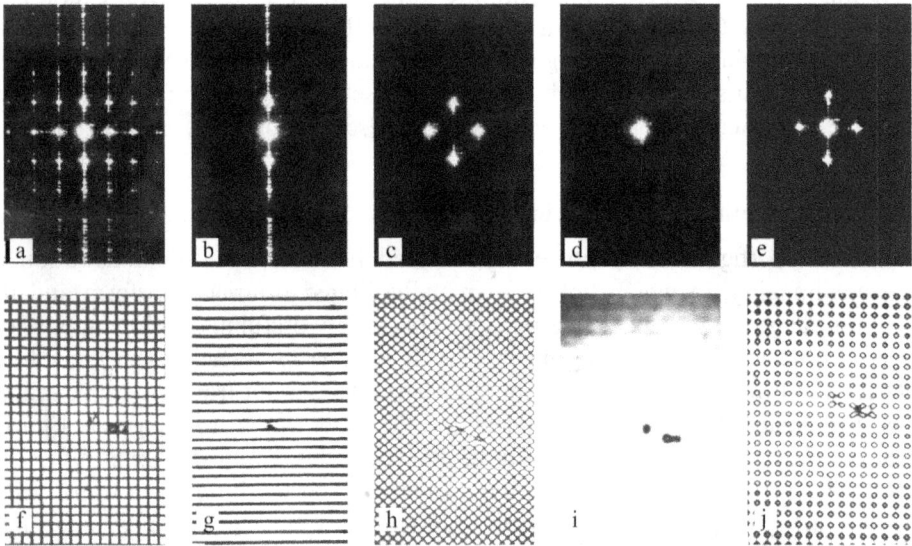

Figure 4.33. Some images obtained when filtering the diffraction pattern in the Fourier plane of a piece of gauze using the system of figure 4.32. (a) is the unfiltered diffraction pattern, and the image of the gauze obtained is shown in (f); the filtered diffraction patterns (b)–(e) are reconstructed as the images (g)–(j). Reconstruction of the zero order (d) alone emphasizes the non-periodic features of the gauze (i). Reconstruction from the zero and first orders only (e) introduces artefacts into the image (j). Reproduced from [2] with permission of Cambridge University Press © 2011.

4.5 Optical transfer function: incoherent resolution measurement

The optical transfer function (OTF, sometimes called modulation or contrast transfer function) describes how an optical system images details with different spatial frequencies. Theoretically, we use an object which has intensity $I_0(x)$ varying sinusoidally with position, with a given spatial frequency $f (= 1/\text{period}$, in units of cycles/mm) and contrast defined by $M_0(f) = (I_{0\,\text{max}} - I_{0\,\text{min}})/(I_{0\,\text{max}} + I_{0\,\text{min}})$. The image is found to be sinusoidal, with contrast $M(f) = (I_{\text{max}} - I_{\text{min}})/(I_{\text{max}} + I_{\text{min}})$. The ratio $\text{OTF}(f) \equiv M(f)/M_0(f)$ is the definition of the OTF. The object can be illuminated by spatially coherent or incoherent light, resulting in different OTFs, but the incoherent OTF is the one of practical importance in imaging instruments. In general, a system will image very low frequencies with high contrast, but as the spatial frequency increases the image contrast gets weaker and eventually reaches zero at the resolution limit. The curve obtained is dependent on the aberrations and other details of the system, and can be shown analytically to be the absolute value of the Fourier transform of the intensity of the point spread function (PSF), which is the image of an ideal point source. It is also dependent on the degree of coherence of the illumination, and it is usually assumed that the illumination is completely incoherent unless otherwise stated (i.e. that the spatial coherence distance in the object plane is smaller than the resolution limit). Coherent imaging resolution was discussed in section 4.4.2. The image of a general incoherently illuminated object with intensity $I(x, y)$ is the convolution of $I(x, y)$ and the PSF, with the appropriate magnification.

Following section 4.1 on Fraunhofer diffraction, the wavefront leaving the aperture stop of the optics has complex amplitude $a(\vec{r}) = |a(\vec{r})|\exp[i\phi(\vec{r})]$, whose Fourier transform is $A(\vec{u}) = |A(\vec{u})|\exp[i\Phi(\vec{u})]$, where $\vec{u} = k_0(\sin \theta_x, \sin \theta_y)$. The PSF is the intensity $|A(\vec{u})|^2$ and its Fourier transform, the incoherent OTF, is therefore the auto-correlation of $a(\vec{r})$, which is the convolution $a(\vec{r}) \times a^*(-\vec{r})$. To get the idea, also try this out analytically in one dimension on a slit aperture

4.5.1 Measuring the OTF using a resolution target

One way of measuring the OTF is to image a resolution target, which contains examples of bar targets with unit contrast at different spatial frequencies, and to analyze the received image of each example in turn to find its contrast. The standard USAF resolution target is shown in figure 4.34.

Another target which contains information at varying spatial frequencies is the radial target used in section 4.2.4, in which the spatial frequency is inversely proportional to the radius (figure 4.35(a)). The example of a *coherent* image in figure 4.35(b) (see also section 4.4.2), shows that the contrast is essentially unity for frequencies up to the resolution limit and zero above it, so that the *coherent* OTF has the form of a step function, falling from 1 to 0 at the resolution frequency limit. Image 4.35(b) shows artefacts which arise from the fact that the object is a square wave (not a sinusoid) and the imaging aperture limits the number of terms in the Fourier series which contribute to the reconstruction of the image. This is known as the 'Gibbs phenomenon'.

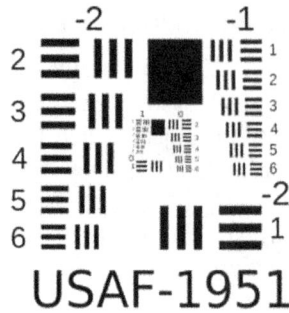

Figure 4.34. Standard USAF resolution target. Reproduced from wikimedia.org, author: Setreset.

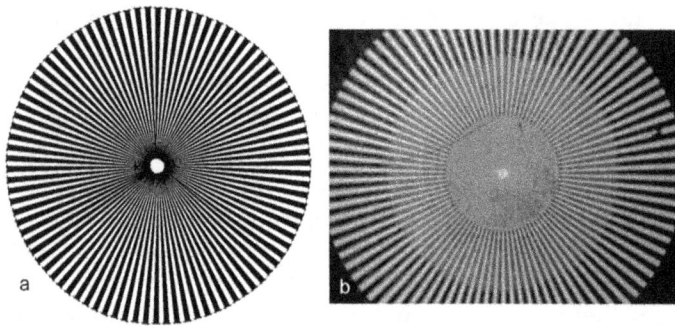

Figure 4.35. (a) Radial target and (b) its coherent image through a lens with finite aperture.

When the same radial target is imaged using incoherent illumination, the contrast varies continuously from the outer edge to the resolution limit, and this limit should be $\lambda/2NA$, which is half the coherent value. An example is shown in figure 4.36. This figure shows two images of the target made with identical optics (see section 4.5.4) using coherent and incoherent illumination, respectively, at the same wavelength.

4.5.2 Random target method

Another method of measuring the OTF used a random target. This target was displayed on a computer screen as a random matrix of 1's and 0's with high resolution (greater than that of the optical imaging system). The display, which is obviously incoherent, was then imaged onto a camera sensor, again having sufficiently high resolution to exceed that of the system. The OTF was then calculated as the ratio between the Fourier transform of the image and that of the source (which should be a constant plus noise). To reduce the noise, the experiment was repeated many times using different randomly created targets. The method was demonstrated [13] using an ideal imaging lens obscured by masks so as to provide 'interesting' OTFs.

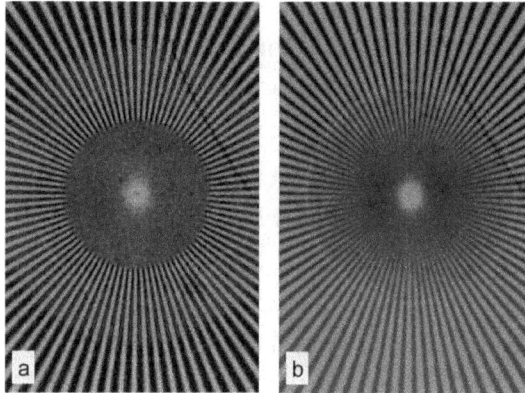

Figure 4.36. Comparison between the images of a radial target obtained with the setup (figure 4.37) described in section 4.5.4 using (a) spatially coherent and (b) incoherent light. Notice that the resolution limit of (b) is better than that of (a).

4.5.3 Using the line and point spread functions

A disadvantage of the methods described above (sections 4.5.1 and 4.5.2) is that they assume that the OTF is independent of the position in the field of view. Other ways of deducing the OTF include analyzing an image of a sharp edge in various orientations and analysing images of point sources. In particular, the use of point sources allows the OTF to be determined as a function of position in the image field, since the function might not be invariant with position. This method is called a 'star test'. For example, if the imaging system has the aberration called 'curvature of field', the image on a plane surface in the paraxial image plane will obviously have a higher resolution on the axis than at substantially off-axis points. The star test gets its name from its use in assessing the quality of telescope mirrors, where a bright but unresolved star is used as the source, and its image is recorded. It is also a widely-used practical method of checking the quality of microscope objectives. In this case, an object is used which is a back-lit opaque slide with several transmitting pinholes at various points in it, the holes being of submicron diameter. A highly magnified image of each pinhole is recorded, and the image is then Fourier-analysed to obtain the OTF in its region of the field. In practice, a typical PSF has a very strong peak surrounded by much weaker details, an image which is difficult to record quantitatively in a single exposure, so that photographs with different exposures have to be taken and combined, which is difficult to do accurately. As a result, when the star test is used by practical lens or mirror polishers it is usual to record the image slightly out of focus so that the variations in intensity are less extreme and more extensive. Such images are widely used as a qualitative indication of the lens quality, but their quantitative analysis is more complicated because the effect of the focus defect, which is a phase error, has to be subtracted from the observed image.

This is an example where the technique of *phase retrieval* [14] can be used to deduce the phase of the wavefront which is incident on the aperture stop (the lens or mirror). It takes into account the fact that this stop has a known shape; it is usually a round aperture of known diameter in which the incident amplitude is constant, possibly obscured by a central obscuration (the secondary mirror of a telescope or reflection microscope). When the phase has been determined this way, the effect of the defocus, which is a quadratic phase change $\phi(r) = ar^2$, can be subtracted out and the remaining phase variations are attributed to the optics. This was one of the methods used to quantify the optical errors in the Hubble space telescope when it was first put in orbit in 1991. Correcting optics (COSTAR) were then designed to cancel the errors and were installed in 1993 [15].

The phase retrieval works in the following manner. First, the amplitude of the observed PSF is calculated (its square root). It is given an arbitrary initial phase, usually either a constant or a random function of position. This is PSF complex amplitude Fourier transformed, and a first estimate of the amplitude and phase of the wavefront in the aperture stop is obtained. The phase of the estimate is retained, but its amplitude is replaced by the known amplitude in the aperture stop, (1 within it and zero outside). This new function is then retransformed to get an estimate of the PSF complex amplitude and phase. Here again the amplitude is replaced by the known measured PSF amplitude, and the phase is retained. The cycle is repeated as many times as needed to get a stable result. After this, the quadratic term $\phi(r) = ar^2$, due to the defocus, is estimated and subtracted from the calculated phase. The OTF is then the auto-correlation function of the obtained complex wave amplitude in the aperture stop. Essentially, this method generates the interferogram which would be achieved in a Twyman–Green interferometer (section 5.2), but it can be very sensitive and is of course independent of the optical quality of the beam-splitter and auxiliary optics in the interferometer.

4.5.4 An OTF lab bench experiment

An experiment which leads to a deeper understanding of the optical transfer functions can be carried out using the slide of figure 4.35(a) imaged through a lens whose aperture can be controlled by an iris diaphragm. We found that a slide with 4° period, i.e. 90 spokes, is ideal for the experiment. The slide is either illuminated coherently by an expanded laser beam or illuminated incoherently by a flashlight with a broadband red filter with centre wavelength about equal to that of the laser, without changing the optics (figure 4.37).

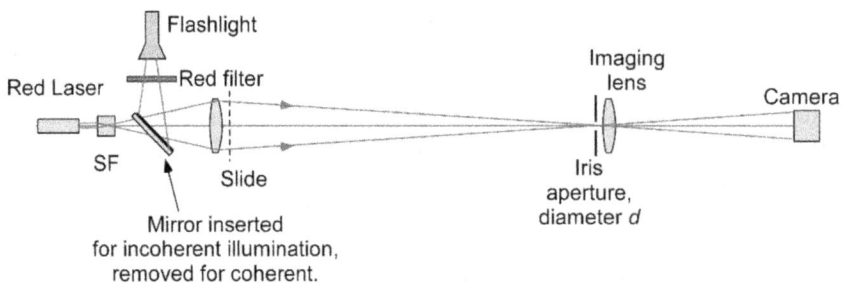

Figure 4.37. Setup used for transfer function experiments.

As in the Abbe–Porter experiment (section 4.4), the lens next to the object slide is adjusted to focus the laser light at the centre of the iris aperture. The camera position is then adjusted with a large iris aperture to get the sharpest image. Now the diagonal mirror is inserted and the incoherent illumination source is arranged to give a uniformly illuminated image of the slide. When the set-up is completed satisfactorily, a series of images, both coherent and incoherent, can be recorded for different aperture diameters. The iris diameter can be determined accurately by closing it onto a drill bit or other rod whose diameter can be measured with a micrometer or calliper after it has been carefully extracted. Actually, since the camera exposure setting may be quite different for the coherent and incoherent images and the iris aperture can be reproduced accurately, it is quickest to take all the coherent images first and then to insert the diagonal mirror and take the incoherent images.

An example comparing the coherent and incoherent images is shown in figure 4.36; notice that the incoherent resolution is better than the coherent, as expected. It is a good exercise in image processing to determine the image contrast of the incoherent image as a function of spatial frequency (radius from the centre of the image), which gives the OTF; the result can be compared to the theory. A further experiment can be to determine the OTF with a measured degree of defocus; in this case, the phase of the OTF is also recorded.

4.6 Diffraction by three-dimensional objects: analogues of crystallography

The importance of the subject of diffraction in general is related to its use in understanding the structures of microscopic objects such as crystals. In particular, the realization that diffraction of x-rays and other short-wavelength radiations could lead to a quantitative measurement of the electron density distribution in crystals and other three-dimensional objects has fuelled numerous fields of physics, chemistry, biology and engineering. Optical diffraction is mainly studied in two dimensions essentially as an analogue to real-life situations using shorter wavelengths, but it is important to understand the influence of the third dimension as well.

Simply put, the Fraunhofer and Fresnel diffraction experiments in this chapter use masks which are a function of position in the (x, y) plane, created on a scale much larger than the wavelength, to diffract light incident in the z-direction. In the case of a two-dimensional mask, the diffraction pattern scales with wavelength, but apart from the scaling there is no other important feature which is wavelength dependent. The question therefore arises: what happens when the object or mask is also a function of the z-direction? Consider the well-known Bragg equation[4], for reflection by an ensemble of parallel planes of atoms in a crystal lattice: $m\lambda = 2d \sin \theta$, where m is an integer. For a given angle of incidence, there will be no reflection unless the ratio λ/d has one of the specific values designated by the equation. So the diffraction

---

[4] In this conventional formulation of Bragg's law, the angle between the axis of the incident wave and the reflecting atomic planes is defined as θ.

depends on the wavelength not just as a scaling factor. One everyday optical example is the interference filter, basically a layered structure which has refractive index which is a function of z. Such filters are designed to transmit a given wavelength, or set of wavelengths, and to reflect others. The important thing to realize is that in this case the diffraction pattern is a non-trivial function of wavelength. Another example, which will be discussed in section 6.2, is the acousto-optic effect in the Bragg regime.

4.6.1 Diffraction by a pair of parallel diffraction gratings: banded spectrum

We will study here a simple example of a structure which illustrates some of the non-trivial features of three-dimensional diffraction. It is a pair of diffraction gratings in parallel planes separated by a variable distance, used to diffract white light. It is an analogy to a crystal with just two lattice planes. The banded spectrum produced by this arrangement was first reported by Barus [16] in 1916, who gave an incorrect explanation for it. The two-grating system was later investigated using quasi-monochromatic light by Lau [17], and interpreted from the point of view of Fresnel diffraction. Further experiments in the field were carried out by Jahns and Lohmann [18].

Let us consider the case of two sinusoidal gratings with period $p \equiv 2\pi/\alpha$, having transmission functions $f(x) = [1 + \cos(\alpha x)]/2$, separated by distance D in the z-direction. When illuminated normally, each grating produces three orders of diffraction: a zero order at angle 0, and 1st and $-$1st orders at angles $\pm\theta_1$ given by $\sin\theta_1 = \pm\lambda/p$. The three orders produced by the first grating are further diffracted by the second grating, each order producing three new orders. There will now be five orders (figure 4.38(a)). There are two second orders at $\sin\theta_2 = \pm2\lambda/p$ resulting from diffraction by both gratings. There is a zero order created by light diffracted by neither grating, in addition to two contributions from light diffracted by one grating in the first order and the other one in the negative first order, and vice versa. Of particular interest to us, there will be two first orders at angles given by $\sin\theta = \pm\lambda/p$ which are each the superposition of two waves, one diffracted into the first order by the first grating and remaining in the zero order of the second grating, and one which was in the zero order of the first grating and was then diffracted by the second. The two waves travel distances $D\cos\theta_1$ and D depending on the directions they were travelling between the two gratings (figure 4.38(b)). They therefore interfere constructively at wavelengths $q\lambda = D(1 - \cos\theta_1)$ ($\approx D\theta_1^2/2$ for paraxial angles) when q is an integer, and interfere destructively when q is an integer plus 1/2. This results in a *banded spectrum*, in which some wavelengths are missing because of destructive interference. The zero order is also modulated as a function of q. The second order is not banded. The important lesson is that *details of the diffraction pattern now depend on the wavelength, not just the scaling.* For example, if we do this experiment using a laser, there will be situations as D changes where the first order is weak, and the zero order is strengthened in compensation.

We could simulate a real crystal by building a periodic stack of many gratings, all separated by D. The result would be constructive interference in the first orders when

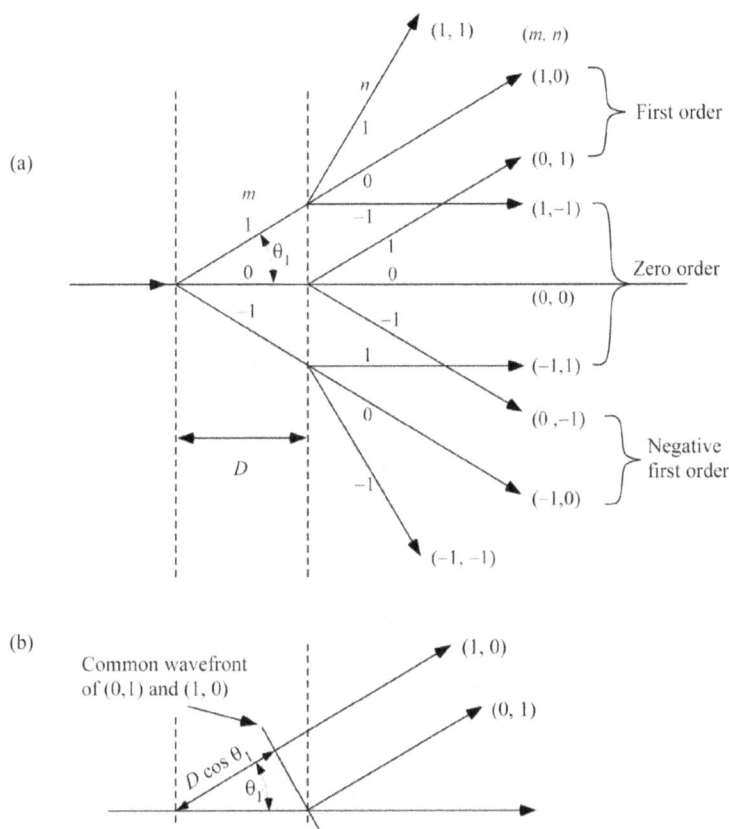

Figure 4.38. Diffraction by the two gratings. (a) The diffraction orders due to the first grating are signified by m, and those of the second grating by n. (b) Interference between the $(1,0)$ and $(0,1)$ orders, with path difference equal to $D(1 - \cos \theta_1)$.

q is exactly an integer (Bragg's law), and zero for all other values, giving sharp diffraction spots at the relevant wavelengths only. But this is not practical on optical scales, although it is commonly done in student labs using microwaves with wavelength several cm.

4.6.2 Carrying out the experiment

The experiment requires us to construct a spectroscope, in which the dispersing element is a pair of identical diffraction gratings mounted in parallel planes in such a way that the distance D between them can be varied (figure 4.39). Using the paraxial approximation above, the bright bands are at wavelengths separated by $\delta\lambda = 2p^2/D$. Using a grating with 300 lines/mm, the minimum distance D must be of order 0.2 mm and the maximum about 10 mm. The light source illuminating the spectrometer slit should be a source emitting a continuous spectrum, i.e. an incandescent lamp, and the telescope optics arranged with a short enough objective lens so that most of the visible spectrum can be photographed in one shot. It is

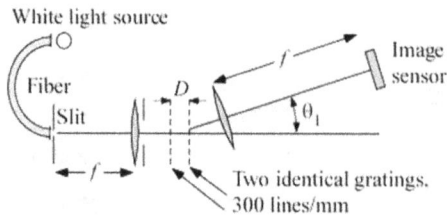

Figure 4.39. Spectroscope layout.

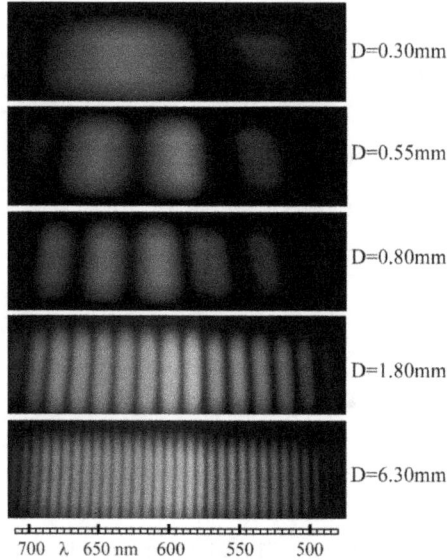

D=0.30mm

D=0.55mm

D=0.80mm

D=1.80mm

D=6.30mm

700 λ 650 nm 600 550 500

Figure 4.40. Some examples of banded spectra.

important to mount the diffraction gratings mutually parallel; an angle between their planes in any direction distorts the spectral pattern. Some examples of banded spectrum made with a pair of 300 line/mm gratings are shown in figure 4.40 for several values of D. In the figure, you can see that the bands are not quite parallel because there must have been a small angle between the gratings in the vertical plane. The gratings used in the experiments shown here were actually blazed gratings, which strengthened the diffracted light in the positive-first orders, but the spectral banding is not dependent on this.

4.6.3 Interpretation in terms of crystal diffraction theory: the Ewald sphere

The relevant theoretical approach to the experiment uses the Ewald sphere construction applied to the reciprocal lattice of the two gratings. A single sinusoidal grating in the plane $z = 0$ is described by the function $f_1(x, z) = \frac{1}{2}[1 + \cos(\alpha x)]\delta(z)$. Its Fourier transform is $F_1(u, v) = \frac{1}{2}\delta(u + \alpha) + \delta(u) + \frac{1}{2}\delta(u - \alpha)$. The second

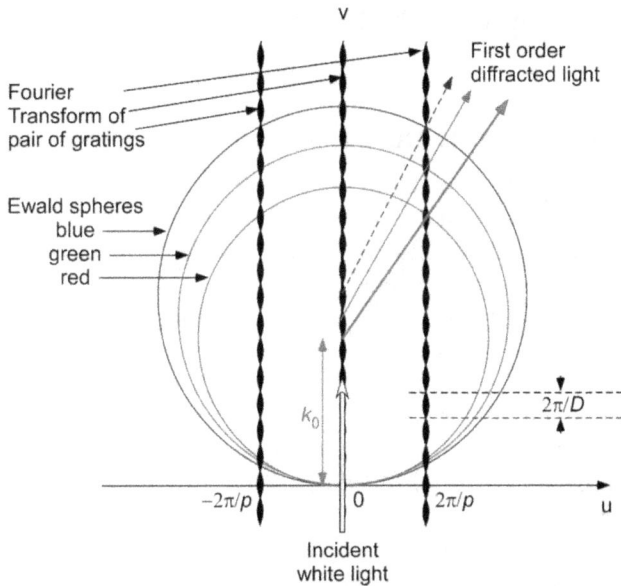

Figure 4.41. Ewald sphere construction in the (u, v) plane showing diffraction by the pair of sinusoidal gratings. Note that the red diffracted vector (from the centre of the relevant sphere) goes through a maximum in the transform, while the blue diffracted vector goes through a zero. The effect of changing the direction of incidence can also be easily visualized by rotating the Ewald spheres about the origin of (u, v).

grating, in plane $z = D$ is $f_2(x, z) = \frac{1}{2}[1 + \cos(\alpha x)]\delta(z - D)$. Its Fourier transform is $F_2(u, v) = [\frac{1}{2}\delta(u + \alpha) + \delta(u) + \frac{1}{2}\delta(u - \alpha)]\cdot\exp ivD$. Their sum is

$$F(u, v) = F_1(u, v) + F_2(u, v) = \left[\frac{1}{2}\delta(u + \alpha) + \delta(u) + \frac{1}{2}\delta(u - \alpha)\right]\cdot(1 + \exp ivD)$$

which has a sinusoidal structure in the v-direction (figure 4.41). The Ewald sphere construction shows that the diffracted light directions are represented by the intersection of this transform with a sphere of radius $k_0 = 2\pi/\lambda$ which passes through the origin and whose centre is determined by the wave-vector of the incident wave. In our case, the incident wave is parallel to the v-axis, and each wavelength has a different value of k_0. This makes clear the origin of the modulation bands.

4.6.4 Interpretation using the Talbot effect

The origin of the banded spectrum can also be attributed to the Talbot effect (section 4.2.3). If we consider the Fresnel diffraction pattern of the first grating in the plane of the second one, there will be certain wavelengths at which a Talbot image will be projected, and if this registers with the second grating, the light will be maximally transmitted. On the other hand, if the conditions are such that the image is displaced by half a period, the light will be blocked. This is another explanation of the phenomenon.

4.7 High resolution, wide field Fourier ptychographic microscopy

Ptychographic microscopy is a recently-developed computational technique which allows a low-magnification microscope to be used to create an image with high-resolution throughout a large field of view, by combining several low-resolution images made under different illumination conditions.

In general, a microscope objective with low magnification can observe a large field of view, but the objective has a low numerical aperture (NA) and so the resolution is limited. This is expressed mathematically as a space-bandwidth product (SBP) [19], which is the number of resolvable points in the image field of view.

In its simplest form, in one dimension, consider a microscope objective with given numerical aperture equal to N and field of view a. In the Fourier plane, this means that the Fourier transform of an object is limited by the lens NA to a region described by $F(u) = \mathrm{rect}(u/k_0 N)$ and the smallest detail is limited by the field size to $\delta u = 2\pi/a$. Thus, the number of distinguishable 'pixels', or degrees of freedom, in the Fourier plane is $2k_0 N/(2\pi/a) = 2Na/\lambda$. This number is the SBP; in two dimensions it is the square of this. Since the number of degrees of freedom in the Fourier plane is equal to that of the image, this is also the number of distinguishable pixels in the image. For commercially available microscope objectives, the field of view is inversely proportional to the magnification because it is limited by a field stop, and the NA is approximately proportional to the magnification of the lens (e.g. an objective ×10 has NA = 0.1; ×100 has NA = 0.9). As a result, the SBP is always of the order of 10 Mpixels; high resolution can only be obtained at the expense of field of view.

The technique of Fourier Ptychographic Microscopy improves the situation by combining several images taken sequentially with the illumination at different angles to the optical axis of the microscope. Each illumination angle projects the Fourier transform of the object into the Fourier plane, but the centre point of the transform is at the illumination angle. The region of each transform which is within the NA of the microscope lens is therefore different. The various regions are photographed in the Fourier plane and are subsequently stitched together mathematically to create a much larger contiguous region, which can then be transformed back to give an image with much better resolution than any one of the individual images. The stitching has to be done carefully, so that both amplitude and phase are identical in the overlap regions, which requires the use of the technique of phase retrieval [14] (section 4.5.3).

A basic description of how the technique works is given in reference [20], while more details are in the breakthrough articles [21, 22]. In these article, polychromatic images were made with SBP of order 1 Gpixel using an array of coloured LEDs switched to provide sequential images with an array of centre points in the Fourier plane.

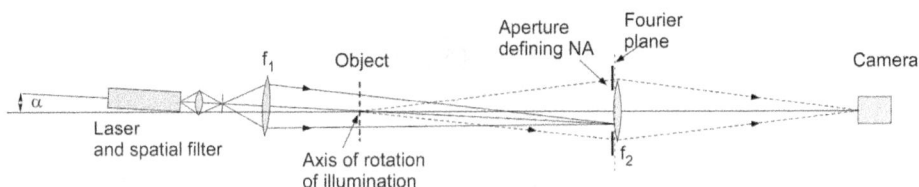

Figure 4.42. Setup used for ptychographic imaging. The angle α was changed in equally-spaced steps.

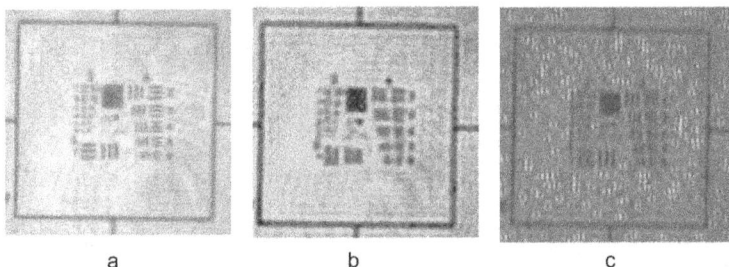

Figure 4.43. (a) High resolution image using maximum lens aperture; (b) low resolution image using on-axis illumination and the small NA; (c) reconstructed image from 15 small NA images at different illumination angles in the horizontal plane. Notice that resolution in the horizontal direction has been improved.

The method was investigated in the lab using monochromatic light from a laser. To understand the technique, it is sufficient to work in one dimension. Similarly to the experiments on coherent imaging in section 4.4, we used the Abbe–Porter system (figure 4.26) and a target (the resolution target in figure 4.34) which was illuminated by an expanded laser beam focused into the Fourier plane by lens f_1. Its image was produced and recorded by a lens f_2 close to the Fourier plane, with a circular iris aperture to define the NA. Next we rotated the illuminating beam mechanically by angle α around a vertical axis through the target, so that the origin of the diffraction pattern in the Fourier plane moved horizontally, and recorded a series of images. It is important that the regions of the transform which pass through the aperture in the Fourier plane and create the successive images overlap substantially (~60%–80%), so that their phases can be matched in the phase retrieval process. The setup we used is shown in figure 4.42. A total of 15 different illumination beams directions were used, at equally spaced angles on both sides of the axis. The improvement in resolution in the horizontal dimension of the final image can be seen (figure 4.43). A major part of the effort was in the programming; see the website [23] regarding this topic.

References

[1] Bragg W L 1975 *The Development of X-ray Analysis* (London: Bell)

[2] Lipson A, Lipson S G and Lipson H 2011 *Optical Physics* 4th edn (Cambridge: Cambridge University Press) ch 7, 8

[3] Taylor C A and Lipson H 1964 *Optical Transforms* (London: Bell)

[4] Harburn G, Taylor C A and Welberry T R 1975 *Atlas of Optical Transforms* (London: Bell)

[5] Senechal M 1995 *Quasicrystals and Geometry* (Cambridge: Cambridge University Press)

[6] Gorkhali S P, Qi J and Crawford G P 2006 Switchable quasi-crystal structures with five-, seven-, and ninefold symmetries *J. Opt. Soc. Am.* B **23** 149–58

[7] Arago F 1819 Rapport fait par M. Arago … au concours pour le prix de la diffraction *Ann. Chim. Phys.* **11** 5–30

[8] Niemann B, Sarafis V and Rudolph D *et al* 1986 X-ray Microscopy with synchrotron radiation at the electron storage ring ar BESSY in Berlin *Nucl. Instrum. Methods* A **246** 675–80

[9] Talbot H F 1836 Facts relating to optical science. No. IV *Lond. Edinb. Philos. Mag. 3rd Ser.* **9** 401

[10] Bakman A, Fishman S, Fink M and Fort E 2019 Observation of the Talbot effect with water waves *Am. J. Phys.* **87** 38

[11] Lord Rayleigh 1881 On copying diffraction gratings and on some phenomenon connected therewith *Philos. Mag.* **11** 195

[12] Porter A B 1906 On the diffraction theory of microscopic vision *Lond. Edinb. Dubl. Philos. Mag. J. Sci.* **11** 154–66

[13] Levy E, Peles D and Opher-Lipson M *et al* 1999 Modulation transfer function of a lens measured with a random target method *Appl. Opt.* **38** 679

[14] Fienup J R 1982 Phase retrieval algorithms: a comparison *Appl. Opt.* **21** 2758

[15] Fienup J R, Marron C J, Schultz T J and Selden J H 1993 Hubble Space Telescope characterization by using phase retrieval algorithms *Appl. Opt.* **32** 1747 (COSTAR)

[16] Barus C 1916 Channeled grating spectra, obtained in successive diffractions *Proc. Natl. Acad. Sci. U. S. A.* **2** 378

[17] Lau E 1948 Interference phenomena in double gratings *Ann. Phys.* **6** 417

[18] Jahns J and Lohmann A W 1979 The Lau effect (a diffraction experiment with incoherent illumination) *Opt. Commun.* **28** 263

[19] Lohmann A W, Dorsch R G, Mendlovic D, Zalevsky Z and Ferreira C 1996 Space–bandwidth product of optical signals and systems *J. Opt. Soc. Am.* A **13** 470–3

[20] Zheng G, Horstmeyer R and Yang C 2014 *Optics and Photonics News* April 2014, p 28

[21] Zheng G, Horstmeyer R and Yang C 2013 Wide field, high resolution Fourier ptychographic microscopy *Nat. Photonics* **7** 739

[22] Ou X, Horstmeyer R, Yang C and Zheng G 2013 Quantitative phase imaging via Fourier ptychographic microscopy *Opt. Lett.* **38** 4845–8

[23] BAIR Berkeley Artificial Intelligence Research https://bair.berkeley.edu/

IOP Publishing

Optics Experiments and Demonstrations for Student Laboratories

Stephen G Lipson

Chapter 5

Physical optics II: interference

5.1 Newton's rings and flat plate interference

5.1.1 Experimental setup

These experiments are the most simple interference experiments and are best carried out under a reflection microscope (figure 5.1), or using the equivalent arrangement set up on an optical bench. This enables the interference patterns to be seen clearly and photographed; a colour camera is recommended. For the most common cases a microscope objective ×2 or ×4 is adequate, or an achromatic lens with focal length 25–50 mm can be used. We should recall, from the start, when the light source is incoherent and extended the interference fringes in both of these experiments are localized in the plane of the optical surfaces; the microscope therefore has to be focused onto this region. Incoherent light from a non-compact lamp or LED source should be used for illumination, not a laser, so that 'accidental' interference between light reflected from irrelevant surfaces, such as the microscope lenses or the second surfaces of the optical elements, is avoided because of the relatively short coherence length of the source. Remember that thin film interference is commonly seen in oil films on a wet road, using the cloudy sky as a source, so there is obviously no need for a spatially coherent source, provided we observe in the plane of localization!

When light is reflected from two optical surfaces in close proximity and separated by air, interference fringes of high contrast are observed, having minimum intensity (close to zero) when the optical path difference is $m\lambda/2$ and maximum intensity when it is $(m + 1/2)\lambda/2$, m being an integer. The reason that destructive, and not constructive, interference is observed when the path difference is a complete number of wavelengths, is because the amplitude reflection coefficients at the two surfaces have opposite signs (section 3.2), one being $(n - 1)/(n + 1)$ and the other $(1 - n)/(1 + n)$. This is also the reason that a soap film becomes black just before it bursts ($m = 0$).

doi:10.1088/978-0-7503-2300-0ch5   5-1   © IOP Publishing Ltd 2020

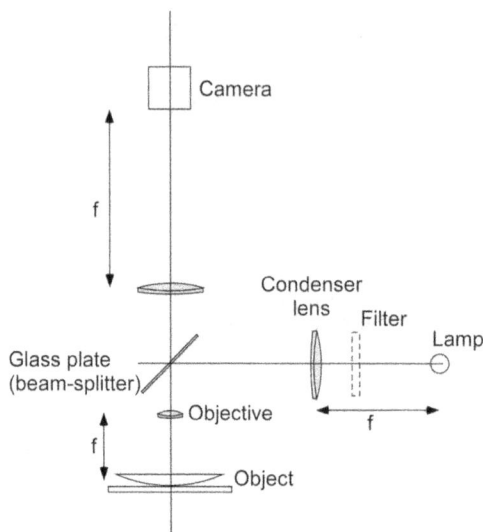

Figure 5.1. Layout of reflection microscope for observing Newton's rings.

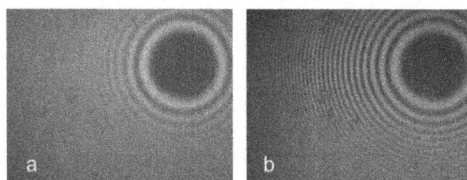

Figure 5.2. Newton's rings in (a) white light and (b) quasi-monochromatic light $\lambda = 508 \pm 5$ nm.

5.1.2 Newton's rings

In the Newton's rings experiment, a singlet lens with positive radius of curvature is placed on a glass plate. Neither needs to be of exceptional quality, but they should be clean and should not be anti-reflection coated. Circular fringes around the point of contact are then observed, the minima being at radii r given by $r^2 = m\rho\lambda$ where ρ is the radius of curvature of the lens surface. What is the focal length of a plano-convex lens appropriate for this experiment? Clearly, when white light is used, the fringes from various colours have different spacings, and cancel one-another out after a few periods (figure 5.2(a)). Introducing a narrow-band filter after the light source increases the number of fringes enormously (figure 5.2(b)). If the central fringe around the point of contact is not dark, the glass surfaces need cleaning!

If a microscope calibration scale is photographed using the same set-up, the radii of the rings can be measured and the radius of curvature of the lens surface deduced. In addition, from the number of fringes observed, the bandwidth of the filter can be estimated; in figure 5.2(b) the number is about 40, so that $\lambda/\delta\lambda \approx 40$, i.e. $\delta\lambda \approx 508/40 = 13$ nm, which agrees roughly with the filter specification ($\delta\lambda \approx 10$ nm).

Figure 5.3. Fringes between two glass plates in the region of contact, with white light illumination.

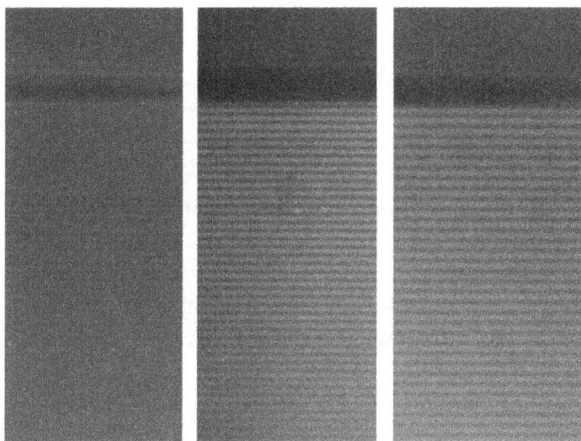

Figure 5.4. The fringes of figure 5.3, when the incident white light is filtered at 486 nm, 540 nm and 676 nm.

5.1.3 Wedge interference

When the Newton's ring system is replaced by two flat glass plates with a small angle between them, straight line fringes are seen. From the fringe spacing, the angle can be deduced. This can be used to measure, for example, the thickness of a hair; in reverse, if the hair thickness is known from a micrometer measurement, the wavelength of the light can be deduced. It is important to confirm that the two glass surfaces do indeed touch, with no space between them; to do this, use a white source (figure 5.3). The accuracy of this deduction should be compared with that of the Michelson interferometer (section 5.2).

Figure 5.4 shows fringes between two glass plates, touching at the top, with white light filtered at 486 nm, 540 nm and 676 nm, in the microscope shown in figure 5.1.

5.2 Michelson and Twyman–Green interferometers: absolute measurement of wavelength, Fourier spectroscopy and optical testing

Albert A Michelson was the father of accurate measurement by means of interference between light waves, which he called 'interferometry' [1]. Many instruments which demonstrated and used interference had been devised before

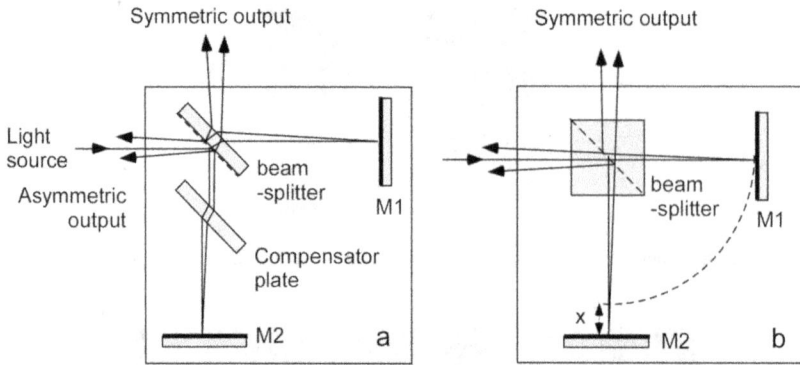

Figure 5.5. Michelson interferometer: (a) original Michelson concept and (b) usual construction today. The rays are shown at a small angle to enable them to be followed; in practice the reflected rays coincide with the incident rays.

Michelson's time [2] (for example, see section 7.1), but the Michelson interferometer, built in about 1880 and shown in figure 5.5(a), is the basic system which has since been used for a wide variety of tasks. Its main advantage over other similar interferometers is that the interfering waves propagate at right angles, and are therefore well separated. Moreover, the output interference pattern propagates in a different direction from the incoming light beam, and is therefore easily accessible. This led to many applications, some of which are as follows:

1. Accurately measuring distances, angles and wavelengths. Michelson received the Nobel prize in 1907 for his application of the instrument to comparing spectral wavelengths with the standard metre.
2. Investigation of the isotropy of space, which led to the special theory of relativity through the Michelson–Morley experiment.
3. Measurement of the quality of optical systems, such as lenses, mirrors and their combinations. The interferometer as used for this purpose is called a Twyman–Green interferometer.
4. Measurement of the temporal coherence function of a light wave, which can be used to determine its spectrum (Fourier spectroscopy). The interferometer as used for this purpose commercially is called FTIR (Fourier transform infra-red).
5. Measurement and detection of small changes in optical path, as has recently been illustrated with the success of the LIGO interferometer in detecting gravitational waves emitted by the mergers of black holes and of neutron stars. This work received the Nobel Prize in 2018.

5.2.1 Michelson's interferometer

The original interferometer was constructed from a plate beam-splitter, a compensating plate identical to the beam-splitter plate, and two plane mirrors M_1 and M_2, whose positions and angles could be controlled delicately (figure 5.5(a)). The purpose of the compensator plate is to equalize the path lengths of the two interfering beams at every wavelength; that is why the plate must be constructed from the same type of glass and have the same thickness as the beam-splitter plate. Today, we usually use a cube beam-splitter B, which in principle needs no compensating plate (figure 5.5(b)). An important consideration regarding the mirrors. We constructed a Michelson interferometer using broad-band dielectric multilayer mirrors, and discovered that the white-light interferogram was quite unexpected in form. Apparently, the effective reflecting surface of a multilayer mirror is wavelength-dependent and this introduces into the interfering waves unknown phase differences, which may vary from mirror to mirror. So it is important to use metallic single surface mirrors, even if their reflectivity is not as good as the dielectric mirrors. This lesson applies also to Fabry–Perot interferometers (section 5.4), for the same reason. Regarding mechanical stability of the interferometer, remember that the interfering light beams travel along different and well-separated paths, and the optical distance between them has to remain constant to better than a wavelength before steady interference fringes can be observed. This means that experiments should be carried out on a stable optical table. Such tables can be purchased, but a simple alternative is a heavy marble slab sitting on three partially inflated tyre inner-tubes. A good way to build a stable interferometer is to bolt the optical elements directly onto a small and thick metal plate, so that the structure is very rigid (section 5.2.3). This construction lacks versatility, but is suitable for several of the experiments.

When discussing the theory of the interferometer, the following points have to be clearly understood.

1. The light source should first be considered as a point source S on the x-axis, which interferes with itself in the symmetric output via two alternative routes. In one route, the wave from S is first reflected at the beamsplitter, then reflected by M_1, and finally transmitted by the beamsplitter. In the second route, the wave is first transmitted by the beamsplitter, then reflected by M_2 and finally reflected by the beamsplitter. In figure 5.6(a) the resulting two images of S are shown when the mirrors M_1 and M_2 are normal to their axes. They are on the y-axis, but at different distances d and $d + x$ from the origin. The interference pattern from these two sources is a set of circular fringes, with their centres on the axis. If we have an extended incoherent source, which can be considered as a collection of independent point sources, each one interferes only with itself and thus will produce a similar pattern, but centred on a different axis, so that for a large enough source the pattern blurs out.

2. On the other hand, (figure 5.6(b)), if the mirrors are at the same distance d from the origin but they are at different angles to the axes, the two images of S lie at the same distance from the intersection between M_1 and the image of M_2 in the beamsplitter, thus on a circle around this intersection. So the fringes are straight lines and the zero order fringe must go through the intersection, wherever the source is situated. This results in fringes which are *localized* in the region of the intersection even if the source is extended. To observe such localized fringes, the camera must be focused on the mirrors' plane. Ideally, the fringes are straight lines in this case, the spacing

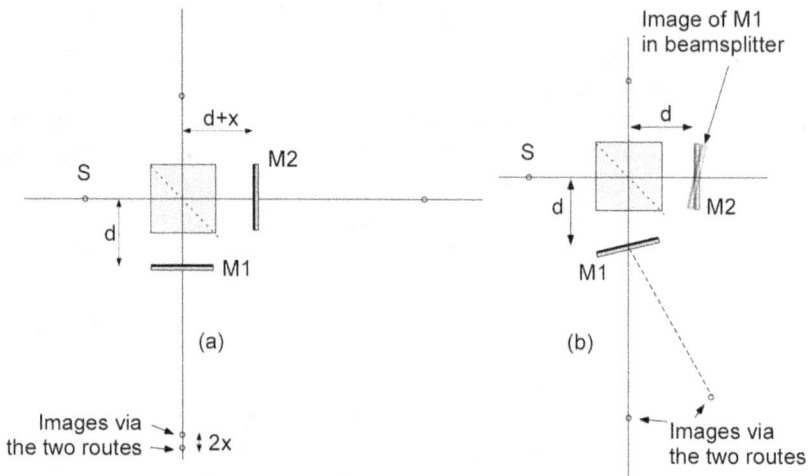

Figure 5.6. (a) Interferometer with M_1 and M_2 normal to the axes, but at distances differing by x from the origin. (b) M_2 is now tilted at an angle to the axis, but its distance from the origin at its centre is equal to that of M_2 ($x = 0$). The image of M_1 in the beamsplitter is shown superimposed on M_2.

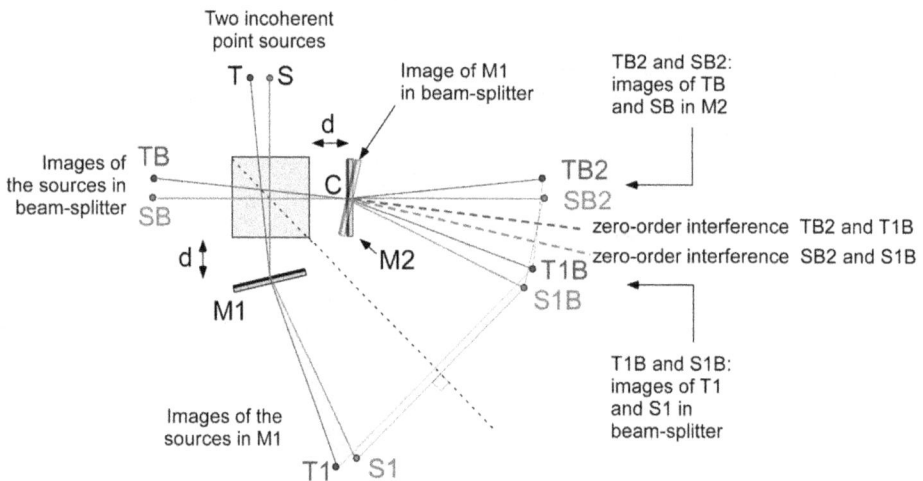

Figure 5.7. Illustrating the localization of fringes when the path difference is zero; the zero orders from all incoherent source points go through C, the intersection between M_2 and the image of M_1. The lines in red and green are not rays; they are geometrical construction lines to show the positions of the images.

depending on the angle θ between M_2 and the image of M_1. Since the path difference is zero along the intersection, the lowest order fringes can be observed with white light. The idea is elaborated in figure 5.7 and examples of white-light fringes are shown in figure 5.8.

3. The reflection coefficients R of the beam-splitter when the incident light is on one side, and $\overline{R}$ from the opposite side, are related by $\overline{R} = -R^*$ where * indicates the complex conjugate. This relationship results from the time-reversal symmetry of Maxwell's equations, and is obvious in the case of

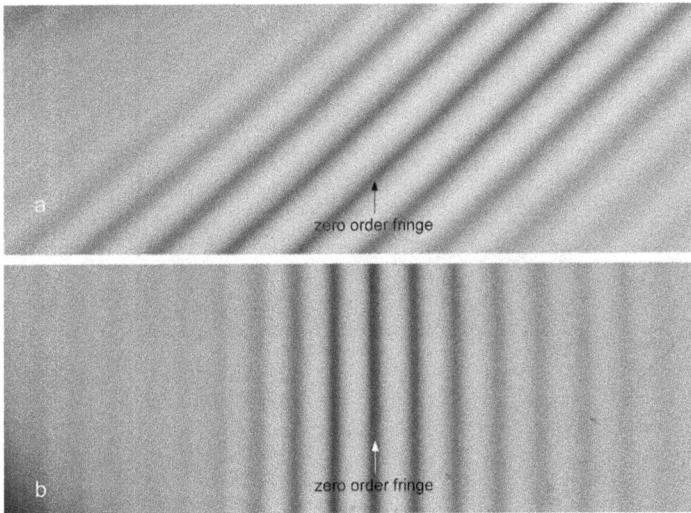

Figure 5.8. Example of white-light fringes, (a) using a white LED source and (b) using a tungsten filament lamp. Note that the zero-order fringe is dark. The fringe patterns are not exactly symmetrical about the zero orders, because the beam-splitter is not ideal.

reflection by a single interface between two dielectrics. If the splitter is symmetrical, R is therefore an imaginary number (i.e. introduces a phase shift of a quarter-wave).

4. There are two outputs from the interferometer, the 'symmetrical' one and the 'asymmetrical' one (see figure 5.5). Because of the relationship $\overline{R} = -R^*$, the interference patterns at these two outputs are complementary, so that one is bright where the other is dark and energy is seen to be conserved (see figure 5.9). The symmetrical output shows interference between waves with amplitudes TR and $\overline{TR}$, where T is the transmission coefficient, which have equal moduli so that the fringes ideally have 100% contrast. On the other hand, the asymmetrical output shows interference between amplitudes R^2 and T^2, which will show less contrast unless $R^2 = T^2$. Fortunately, the symmetrical output is more accessible; this is a very important practical feature of the Michelson interferometer. But the fact that $\overline{R} = -R^*$ means that there is *destructive* interference in the zero order of the symmetrical output, which is therefore dark.

The experiments on the interferometer can be carried out best using a camera in the symmetrical output, with a lens focused on the interferometer mirrors where the fringes are localized. A colour camera is advisable, because the experiments are not limited to monochromatic light (figure 5.10).

The interferometer should be constructed on the optical table and aligned visually using a laser, first using the unexpanded laser beam and then after it has been expanded. We will assume that mirror M_1 is fixed in position, but its angle θ can be changed, whereas M_2 is always normal to the laser beam, but its position x can be translated along the optical axis. When $x = 0$ and $\theta = 0$, one mirror coincides with

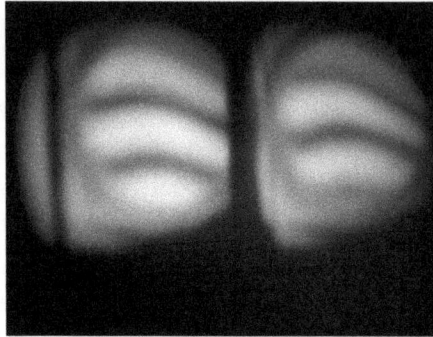

Figure 5.9. Complementary patterns in the anti-symmetric (left) and symmetric outputs (right). Note that the zero-order fringe on the right is not quite black; the beam-splitter is apparently not ideal. Nor are the fringes straight, which tells us something about the quality of the optical elements used!

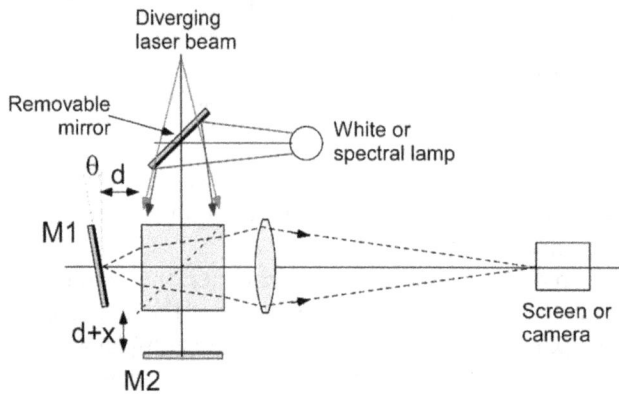

Figure 5.10. Layout of experimental arrangement which allows both laser and other light sources to be used.

the image of the other mirror in the beamsplitter—but these conditions have to be determined experimentally.

5.2.2 Fringe types in interferometers

Using the expanded laser source, the different types of fringes obtainable should be investigated. First, if the laser beam is focused to a point and then expands as a spherical wave, and x and θ are arbitrary, we see curved interference fringes. By adjusting θ about both vertical and horizontal axes, the fringes can become circular, with their centre in the field of view. As x approaches zero, the separation of the fringes becomes greater, until the field becomes uniform when $x = 0$. When this condition is achieved, changing θ gives rise to parallel straight fringes. These experiments can be done quantitatively, measuring for example the radii of the rings as a function of x, or the separation of the straight-line fringes as a function of θ.

5.2.3 Measuring the wavelength

Michelson [1] used the diverging-wave configuration (circular fringes) to measure the wavelength of a Cd emission line known to have no fine structure, by counting fringes at the centre of the pattern as x was changed by calibrated distances. This is incredibly difficult to do by hand if the interferometer is built on an optical table using standard post mounts for the mirrors and beamsplitter, since the fringes jump whenever the setup is touched. However, if the mirrors and their mounts are bolted directly onto a rigid base plate, as Michelson did, it is possible to overcome this problem. Another possible method is to move one mirror by means of a lever with a known mechanical ratio, operated by a micrometer detached from the optical table; we tried this but we found it not to be accurate because the lever bent slightly. If a computer-operated translation stage is available, it can be used to move one mirror steadily enough for the fringes to be counted. In this case, the calibration depends on the manufacturer's specification for the translation motor. It is still easy to miscount the fringes, so it is advised to take a video-recording of the interference pattern as the mirror is moved, so that the accuracy of the counting can be checked in retrospect.

An alternative method of measuring the wavelength is to measure the fringe spacing in the symmetrical output as a function of the angle θ between the interfering waves, measured using the asymmetrical output (figure 5.11). When the laser beam is collimated by addition of a converging lens, the fringes will always be parallel and straight for any x and θ, and have separation $\lambda/[2 \sin(\theta/2)] \approx \lambda/\theta$ when θ is small. It is important that the straightness of the fringes be confirmed first at the largest θ used, to ensure that the beams are exactly collimated. The linear scale of the camera sensor has to be calibrated with a reticle in contact with it, illuminated by a collimated beam, and the angle θ can be measured by extracting the two beams in

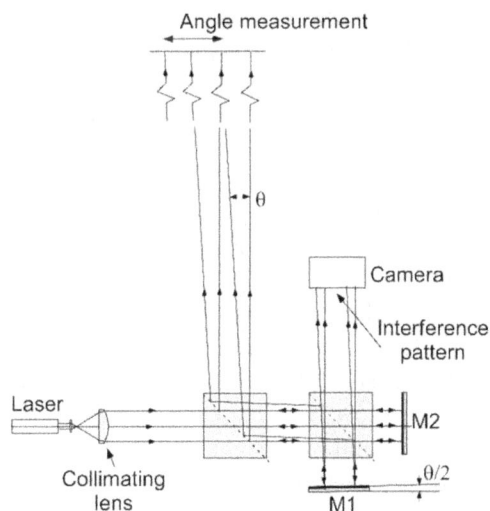

Figure 5.11. Setup for measuring the wavelength using parallel fringes.

the asymmetrical exit and projecting them on a screen at a large known distance; then θ = separation of the beams/the distance from the mirrors. The standard of length to which the wavelength is compared in this case is the reticle, since angle is dimensionless. Note that since the reticle illumination is collimated, and the reticle is separated from the sensor by a short distance, the image of the reticle may not be an exact contact image because of Fresnel diffraction, but its period is still equal to that of the reticle. It is an interesting exercise to estimate how accurate this method can be, taking into account that to see the fringes clearly their spacing must be at least 4 pixels, although the other dimension of the pixel array might also be used to picture them more clearly if the fringes are not parallel to either dimension.

Some points to be re-emphasized are as follows. The fringe spacing is related to the angle between the two wavefronts which are intersecting. If the wavefronts are not planar and the two wavefronts do not exactly coincide, the local angle may differ from the average angle of propagation of the beam, and the fringes may not be equally spaced. It is therefore important that the wavefronts be exactly planar. This requires the input beam to be collimated and of somewhat greater diameter than the region of intersection with the camera, so that possible curvature of the wavefronts at the edge of the beam is outside the region of interest. The degree of curvature of the wavefront which is allowable should be such that there is less than one wavelength deviation from a plane over the region of interest. If this has diameter D (the diagonal of the camera sensor) this means that one must ensure that the beam does not come to a focus within a distance of at least $D^2/8\lambda$ (~12 m).

5.2.4 White-light fringes and spectroscopy

Once the interferometer is assembled and adjusted to give about 10 *straight* fringes across the field of view with x close to zero, it can be used with incoherent light and for spectroscopy. Remember that the lens is needed to image the localized fringes on the camera (figure 5.10). If the laser is then replaced by a white light source (an LED flashlight is good), it may be possible with luck to catch the white-light fringes by adjusting the position of M_2 in small steps till the path difference x is zero. Then, one should note that the achromatic zero-path-difference fringe is dark, not bright. A more systematic method to find the zero position is first to use a Na discharge lamp as the source. The doublet at 589.0 and 589.6 nm gives rise to fringes with a beat modulation period Δx of about 1000 fringes, and these can be seen even when $|x|$ is of order 1 mm. Then, by changing x in the direction in which the beats become more pronounced until the highest contrast is obtained, the white-light fringes can usually be found immediately in that locality. It is interesting to devise a way in which to observe simultaneously both the symmetric and anti-symmetric outputs of the interferometer, and to show that the pictures are complementary (figure 5.9). It is also a useful idea to include an adjustable polarizer before the camera to improve the contrast, since the beam-splitter might behave differently in the p- and s-polarizations.

5.2.5 Fourier spectroscopy

Observing the beats Δx between the two wavelengths of the Na doublet is the first step towards Fourier spectroscopy. Other sources can be used and the fringes measured as a function of x using the camera. Fourier spectroscopy was actually invented by Michelson and is described in his 1927 book [1], but he could not put it into practice in a serious way because he did not have a computer available. The method normally requires measuring the interference intensity at the centre of a circular fringe pattern as the displacement x is increased in a controlled manner. This is very difficult to do with a system mounted on an optical table, because of its sensitivity to being touched, but the use of a computer-controlled translation stage can solve the problem. An alternative method using a camera is as follows. A dense pattern of linear fringes (parallel to z) is displayed by the camera (y–z plane) for a given angle of tilt on mirror M_1. Then each position y on the mirror is equivalent to a given value of $x = 2\theta y$. The fringe profile can be measured from the recorded photograph, or better from a series of overlapping photographs as x is changed in steps. Following this, the profile can be Fourier transformed to generate the spectrum of the source. If the camera has, say, 4000 pixels along the y-axis, it is feasible to record 1000 fringes along this axis, so that a resolution of $\lambda/\delta\lambda = 1000$ can be obtained from a single photograph; not really very high resolution, but sufficient to illustrate the idea. For example, a LED might have a bandwidth of 5 nm, so that its spectral profile can easily be measured this way. Or the transmission spectrum of a narrow-band filter might be measured. The spectrum of a white LED flashlight is also an interesting experiment, in which it should be compared with that of a white filament lamp (see figure 5.8). A question to be discussed: if the beamsplitter is not ideal, so that the white-light interferogram is not symmetrical about the zero order fringe (as in figure 5.8), does this affect the accuracy of Fourier spectroscopy?

5.2.6 Optical testing—the Twyman–Green interferometer

In the Twyman–Green form of the interferometer, one mirror is replaced by an optical system to be tested (figure 5.12). The source of illumination is usually

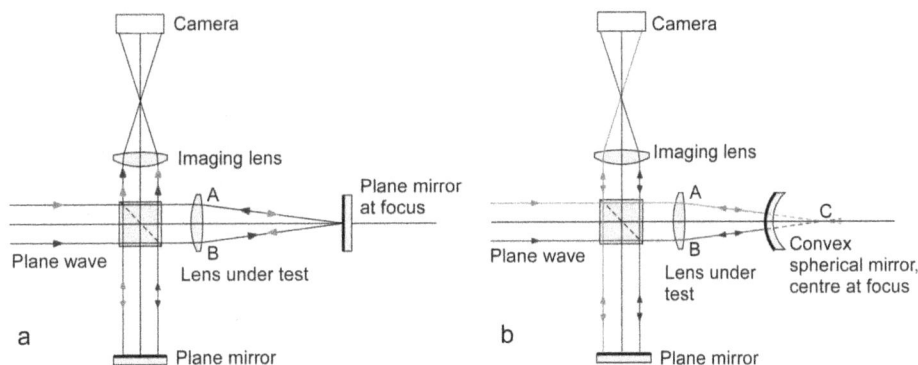

Figure 5.12. Two variants of the Twyman–Green interferometer used for testing the quality of a converging lens.

monochromatic, i.e. a laser. The tested system has to be designed so that if it is illuminated by a plane wave, a plane wave is returned if the design is perfect, and this returning wave interferes with the reference wave in the other channel, which is reflected from the plane mirror. This is not difficult to do if the interferometer is used for testing a plane mirror or a window or a prism with plane surfaces. But it is not quite as simple when testing a lens or lens combination, because in order to get quantitative information about errors in it one has to make sure that each point on the interferogram corresponds to a specific point in the aperture stop, and since the light passes twice through the system, it may be that it samples two different points. For example, consider the system shown in figure 5.12(a), where a plane mirror is situated at the focal point of the lens under test in order to return the rays. If the lens is ideal one indeed returns a plane wave to interfere with the reference wave. But it is clear from the figure that the returning ray has passed through the lens at two different points A and B, so that any wavefront error observed cannot be attributed to a single point. On the other hand, if the plane mirror is replaced by a spherical mirror with its centre at the focal point of the lens (figure 5.12(b)), each returning ray passes twice through the same point on the lens (or the same region if the lens is not ideal) so that errors can be attributed to a single point or region of the lens. Of course, setting up this second system accurately is more difficult, but the output is more useful to the lens producer.

5.2.7 Interpreting interferograms quantitatively

When an intererogram has been recorded, the next stage is to interpret it so as to obtain useful information about the tested system. This requires converting the interference pattern to a picture indicating phase as a function of position in the field. The amplitude is usually less relevant. One way of doing this is to use a 'phase shift algorithm' which derives the phase φ (modulo 2π) at each point from three or four sample interferograms taken with the plane mirror moved in steps of $\lambda/8$ (additional phase change of $\pi/2$ in each successive picture). In the case of four samples, the intensities are $I_1 = I_0 + I \sin \varphi$, $I_2 = I_0 + I \sin(\varphi + \pi/2) = I_0 + I \cos \varphi$, etc, where I_0 is the mean background intensity and I is the modulation amplitude. It is easy to show that $\tan\varphi = (I_1 - I_3)/(I_2 - I_4)$ from which φ mod 2π at each point can be calculated. A similar result can be found using only three steps. In practice, the four (or three) pictures can be found if a video recording is made of the interferogram while a slight pressure is applied to the plane mirror mount; watching any given point as it goes through a full cycle allows four frames separated by $\pi/2$ to be selected. Finally, it is necessary to find the value of φ at each point absolutely, which eliminates the phase jumps of 2π, making φ a continuous function. This requires a two-dimensional phase unwrapping algorithm, of which many are available.

5.3 Sagnac common-path interferometer

The Sagnac interferometer is a version of the Michelson interferometer in which the two interfering beams travel along a common path (figure 5.13). Since the paths are identical, environmental changes such as vibrations or thermal air currents affect

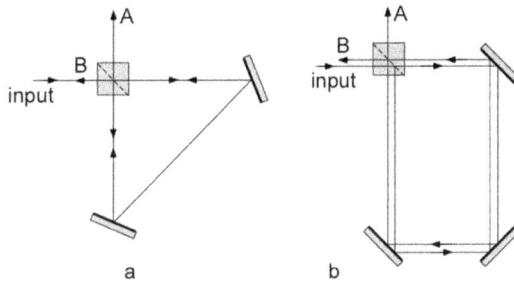

Figure 5.13. Two versions of the Sagnac interferometer. A is the asymmetric output, and B is the symmetric (preferred) output. (a) Triangular Sagnac, (b) rectangular Sagnac. In (b), a small translation of the beamsplitter causes the counter-rotating beams to overlap exactly.

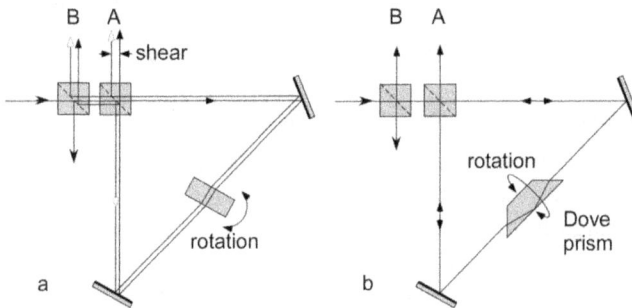

Figure 5.14. Triangular Sagnac arranged as a shearing interferometer. In (a) the shear is obtained by rotating a thick glass plate situated in the common path about an axis normal to the plane, and in (b) rotational shear around the light beam axis is obtained by inserting a Dove prism and rotating it about the common path.

both beams equally, and so the interference pattern is very stable; you can even use it in a helicopter! For this reason, I am a great fan of the Sagnac, and use it whenever possible for interferometric experiments. Two versions are shown in the figure; in the first one, the beams travel exactly the same path, whereas in the second one, the paths can be slightly different, so that it is possible to modulate them independently, while still retaining the advantages of mechanical and environmental stability. A minor disadvantage of the triangular setup, when compared with the Michelson interferometer, is that the symmetrical output, which gives high contrast fringes even if the beam-splitter is not 50% reflecting, returns in the direction of the source. However, this can be overcome by employing a second beamsplitter (as in figure 5.14). The most important difference between the two paths is that the light travels one path in a clockwise sense and the other anticlockwise. As a result, the output is sensitive to rotation of the frame of reference, and this leads to the best-known application of the interferometer, which is the optical gyroscope. This topic is discussed in section 9.2 on relativistic optics, since even for small rates of rotation, special relativity is needed to describe its behaviour in a moving frame of reference. Because the path lengths are automatically almost exactly equal, interference fringes in the Sagnac are easily obtained using white light. In several of the applications mentioned later, the beam which travels through the interferometer is broad and can contain an image or other information, in which both the amplitude and phase are at each point are relevant. When a broad incoherent source is used, the interference fringes created are localized at infinity.

5.3.1 Aligning the interferometer

The interferometer should initially be aligned using a laser. First the mirrors are arranged *without* the beam-splitter, so that the output beam A (clockwise) goes in the required direction and intersects the incoming beam at the same height. Then the beam-splitter is inserted at the intersection, and *by adjusting it alone* the anticlockwise output is arranged to overlap the clockwise one. Then interference fringes are immediately visible, and final adjustments can be made, still using the beam-splitter alone.

5.3.2 Sagnac interferometer in a stationary frame of reference

In the laboratory, the interferometer has applications based on its exceptional stability. The question with common-path interferometers is how to modulate the counter-propagating beams independently. In the triangular interferometer (figure 5.13(a)), inserting a glass plate in the beams at a variable angle, allows the exiting beams to be shifted relatively (figure 5.14(a)). This creates a shearing interferometer, in which we can see interference between different spatially-separated parts of an extended wavefront, and can be used to measure spatial coherence, for example (section 6.4), or the phase gradient of a wavefront. In the former example, the coherence function is directly related to the contrast of the fringes observed, as a function of the shear. Alternatively, a Dove prism can be inserted to provide rotational shear [3] (figure 5.14(b)), which can be used to measure the spatial coherence of a uniform plane wave 'in one shot', since at each radius, the shear is different. In order to measure the fringe contrast accurately, it is preferable to use the symmetric output (B in figure 5.13) since the fringe pattern there lies on a negligible background, and this is most easily done by incorporating a second beamsplitter as shown, even though it wastes three-quarters of the input light intensity.

In figure 5.15(a), we see a variant of the rectangular Sagnac, in which shifting the input beam laterally separates the two counterpropagating beams, so that transparencies or cells in the two beams can be compared inteferometrically. Because the set-up is almost common path, it works well in white light; the addition of differential path-length correctors (figure 5.15(b)) allows the path lengths to be equalized exactly, or to introduce a fixed path-difference in order to compensate a substrate, for example. Additionally, with this set-up there is no problem in observing the symmetric output, since it is displaced laterally from the input beam, and can be picked up with a mirror. Figure 5.16 shows both A and B outputs photographed simultaneously.

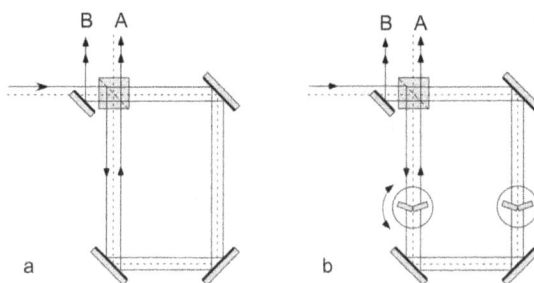

Figure 5.15. The rectangular Sagnac interferometer arranged to provide (a) interference between separated beams traveling in opposite senses and (b) with differential path-length adjustment.

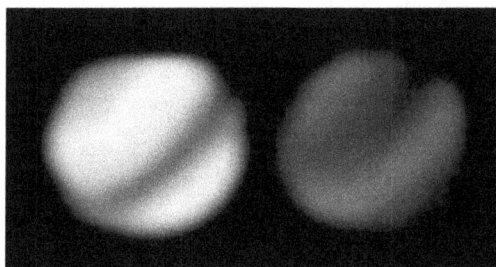

Figure 5.16. An example of the B and A outputs photographed together in the setup of figure 5.15(b). Note the better contrast in the B output.

5.3.3 Fourier spectroscopy with a Sagnac interferometer

Another application of a stationary Sagnac is to make a low-resolution Fourier spectrometer with no moving parts. Essentially, the virtue of the Sagnac here is simply in its stability and the ease with which zero path difference can be obtained. We use the configuration of figure 5.15(b). Light from source with an unknown spectrum (e.g. a star) is focused by a lens into the system, after the interferometer has been aligned using a laser. At the output (preferably the B output) there is a second lens which forms an image of the first one onto the camera sensor. The interferometer is then adjusted to give a dense fringe pattern with the zero-order fringe at the centre, or at one side of the pattern. It is not easy to get such a pattern by just adjusting the mirrors, because they affect both beams, but the beamsplitter can be used for this.

The interference pattern observed consists of a superposition of the interference fringes created by different wavelengths in the source, which can be Fourier-analyzed to provide the spectrum (Fourier spectroscopy, section 5.2.5). The pattern should be symmetrical about the zero-order fringe and its transform is a real function, as a spectrum should be. However, optical aberrations in the system may introduce phase errors, and the pattern might not be symmetrical, but still the intensity $|F(k)|^2$ of the Fourier transform represents the spectral intensity.

The resolution of the spectrometer is determined by the number of fringes photographed, and this in turn is limited by the Nyquist condition that at least four pixels must sample one period of the interferogram. This means that if the length of the camera sensor is N pixels, the largest number of fringes is $N/4$ and so the resolving power is $\frac{\lambda}{\delta\lambda} = \frac{N}{4}$. If $N = 5000$, for example, the resolving power is 1250 which, for a spectrometer, is not very high resolution; compare this with a diffraction grating (section 4.3) or Fabry–Perot (section 5.4)! This value can be improved by using the two axes of the camera sensor, when the fringes are diagonal. It is an interesting exercise to find the optimum configuration. Take into account that the optics are not perfect, so the fringes may not be straight; but the configuration can be calibrated using a laser. Incidentally, for spectrometry, the Sagnac configuration has no particular advantages over other interferometers except for stability and white-light capabilities, and other interferometers like the Mach–Zehnder or Jamin might replace it [4]. Another way in which the Sagnac interferometer can be used for low-resolution spectroscopy is described by Malik *et al* [5] and Schröck *et al* [6].

5.3.4 Optical testing using the Sagnac interferometer

In the form of figure 5.15(b), the Sagnac can be used for phase imaging of a transparent object. The object is placed in one of the circulating beams and a parallel-sided plate window of approximately the same thickness in the other one. The interference pattern then indicates the phase difference between the two. The differential path-length adjustment plates can be used to bring the average path-difference to zero, and then small path differences from point to point will appear in different colours if white light is used for illumination (figure 5.17). Otherwise, using quasi-monochromatic illumination (better than a laser because speckle is avoided) an interferogram is formed which can be interpreted quantitatively.

The great advantages of the Sagnac for this purpose are first that the zero path difference can be achieved very easily, and second that it can be operated in an environment which is not mechnically stable.

5.4 Fabry–Perot étalon

The Fabry–Perot étalon is a multiple-beam interferometer with exceptionally high resolution [7, 8]. Its construction is very simple, since it consists of two parallel plane mirrors with high, but not unit, reflectivity $R = 1 - \Delta$. Most of the incident light

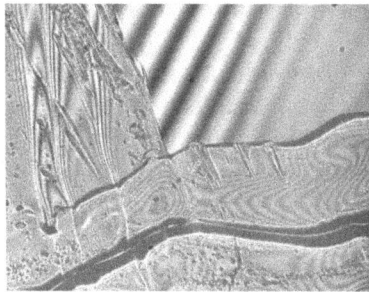

Figure 5.17. Phase image of a transparent object, using white light illumination in a Sagnac interferometer of the type shown in figure 5.15(b).

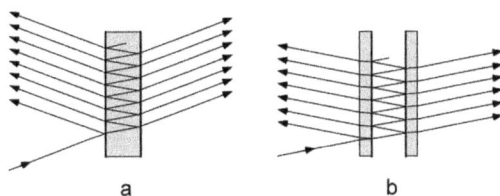

Figure 5.18. Two constructions of the Fabry–Perot étalon.

which is transmitted by the first mirror is reflected back and forth between the mirrors, but a small amount is transmitted on each reflection and the escaping waves interfere (figure 5.18). The intensity after p reflections is proportional to $R^p = (1 - \Delta)^p \approx e^{-p\Delta}$, which means that the intensity is reduced to $1/e$ after $1/\Delta$ reflections. Using the general interferometric result for the resolving power, $\lambda/\delta\lambda = Nm$, where N is the number of interfering waves and m is the order of the interference (number of wave periods between successive waves), we can immediately expect that the resolution is of the order of $2d/(\lambda\Delta)$. From an exact calculation, we find that the width of an interference fringe at half maximum intensity is equal to the fringe spacing divided by the 'finesse' which is defined as $F \equiv \pi\sqrt{R}/(1 - R) \cong \frac{\pi}{\Delta}$. For example, if the separation of the mirrors is 10 mm, and the reflection coefficient of each mirror is 98%, we get a finesse of 150 and resolution of order 6×10^6. Constructive interference between the waves transmitted by the étalon occurs when $2nd \cos\theta = m\lambda$, for integer m, where θ is the angle between the wave-vector and the normal to the mirrors within the étalon space, d is the spacing and n is the refractive index within the space. Since only a fraction Δ of the incident light seems to be transmitted by the first mirror, it is surprising that when the interference of the waves transmitted by the étalon is constructive, all the incident light is transmitted and none is reflected; this occurs because under the same conditions, the waves returning in the direction of the source all interfere destructively. Figure 5.20 shows a typical interference pattern obtained with laser light emanating from a point focus and finesse 8 (reflection coefficient 68%).

5.4.1 Laboratory model

A simple Fabry–Perot can easily be constructed either (a) by coating the two faces of a thick parallel-sided glass window with semi-reflecting metallic layers (figure 5.18(a)), or (b) by using two semi-reflecting metallic mirrors mounted face to face in a manner that allows the angle and distance between them to be adjusted with high sensitivity (figure 5.18(b)). To achieve the theoretical resolution, the quality of the flatness and polishing of the plates must be very good. We should point out again that the mirrors should have metallic coatings, and not be broad-band multilayer dielectric mirrors (see section 5.2.1); in a multilayer reflector the effective plane at which the reflection occurs is not uniquely defined, and may be wavelength or angle dependent.

The advantage of construction (a) is that the étalon is rigid and well-defined. However, its parameters cannot be changed—in particular, the plate has to be chosen so that the two surfaces are accurately parallel. A simple method of constructing such

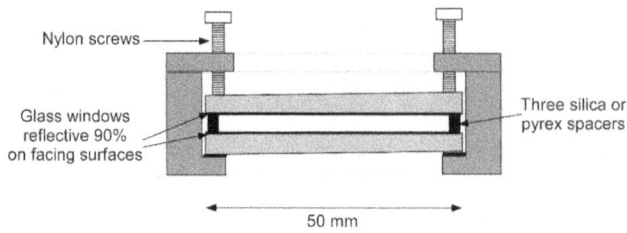

Figure 5.19. An étalon constructed from two optically flat windows.

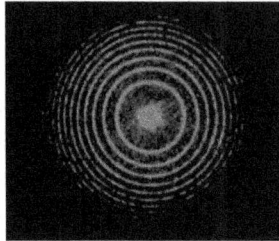

Figure 5.20. Interference pattern obtained with the type (a) étalon described. Reproduced from [9] with permission of Cambridge University Press © 2011.

an étalon is to use a piece of high quality mica, cleaved carefully so as to have no steps on its surface within a region of about 1cm square. It can then be coated with aluminium. Clearly, its surfaces are atomically flat and parallel, but it is difficult to find high quality mica more than a fraction of a mm thick. One should also be aware that mica is biaxially birefringent, so that unpolarized light gives rise to two superimposed fringe patterns with orthogonal polarizations because of the two values of n, and this may be mistaken for the pattern from a spectroscopic doublet!

Construction (b) allows the separation to be changed for experimental purposes. To achieve the theoretical resolving power, it is necessary to have surfaces which are polished to a high degree of accuracy. For example, if Δ is 20% there are about 5 reflections which have to interfere constructively, so that random surface errors should not be greater than $\lambda/10$. Figure 5.19 shows a design for construction (b) which is not difficult to implement. It uses two optically flat windows, coated with Al to have a reflection coefficient of about 90%. The spacers should be three small blocks or discs cut adjacently from the same parallel-sided plate of a material with a low thermal expansion coefficient.

5.4.2 Interference pattern

To investigate the properties of the étalon, we take into account the fact that the condition for constructive interference is dependent on the wavelength λ, the optical path difference and the angle of propagation between the mirrors i.e. $2nd \cos \theta = m\lambda$. Thus, an extended source can be used and an observing telescope or camera focused on infinity, so that each point in the observed field corresponds to a given angle θ. An interference fringe corresponds to constant $\cos \theta$, which defines a cone in space, or a circle in the image plane; thus we see a set of sharp circular fringes superimposed on the image of the extended source. The centre of the fringe pattern corresponds to the highest order, $m_0 = 2nd/\lambda$, which may not be an integer. Figure 5.20 shows fringes from an étalon of type (a) constructed from a standard optically flat (0.25λ) glass window 5 mm thick coated with Al films 85 nm thick on each side. Its resolving power, estimated from the width of the rings relative to their spacing (~5) and the plate thickness, is of order 10^5. The profile of the fringes in figure 5.20 is compared with the expected profile for ideal reflectors with finesse 8 in figure 5.21.

5.4.3 Measuring the thickness of the étalon

In a quantitative experiment, it is necessary to determine the optical thickness of the étalon. One method, which can be accurate to about 1%, uses a monochromatic source with a well-known wavelength. A measuring telescope is used for observation (figure 5.22), which can be rotated on an angle scale till the cross-wires coincide with successive fringes, and the fringe number plotted versus $\cos\theta \approx (1 - \theta^2/2)$; the gradient then gives nd (figure 5.23). Alternatively, a single photograph of the fringe pattern can be used, in which case it is necessary to know the dimensions of the sensor and the focal length of the imaging lens.

A much more accurate method for an air-spaced étalon (figures 5.18(b) and 5.19), devised by Benoit who worked together with Fabry and Perot, uses several (3 or 4) accurately-known wavelengths. The plot of figure 5.23 is then used to determine the fractional part of the central fringe order for each wavelength. One then finds the full values of m_0 for each wavelength, such that the fractional parts are consistent at the

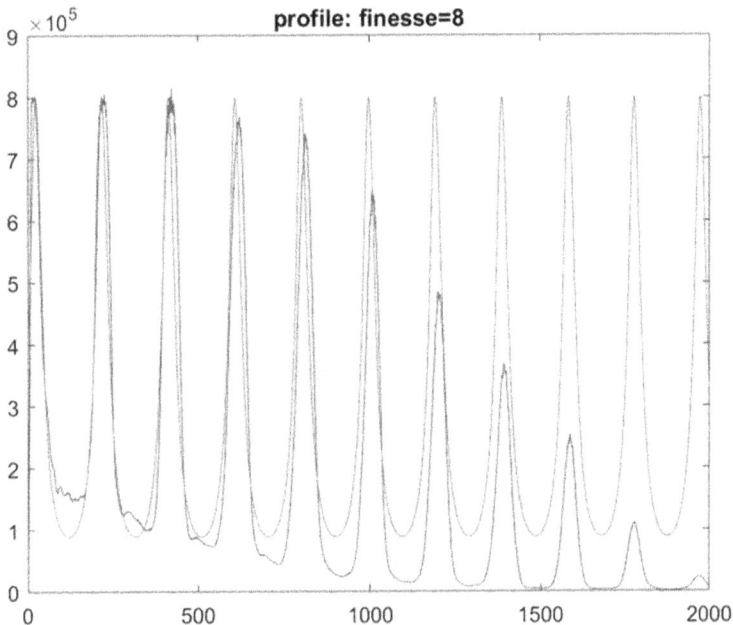

Figure 5.21. Measured fringe profile of figure 5.20 (blue) compared with expected profile for finesse = 8 (orange).

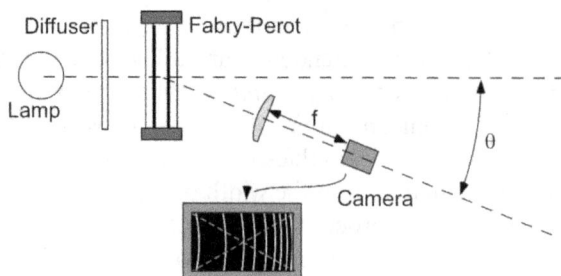

Figure 5.22. Schematic layout of the experiment. The cross-wires on the display are used to determine θ.

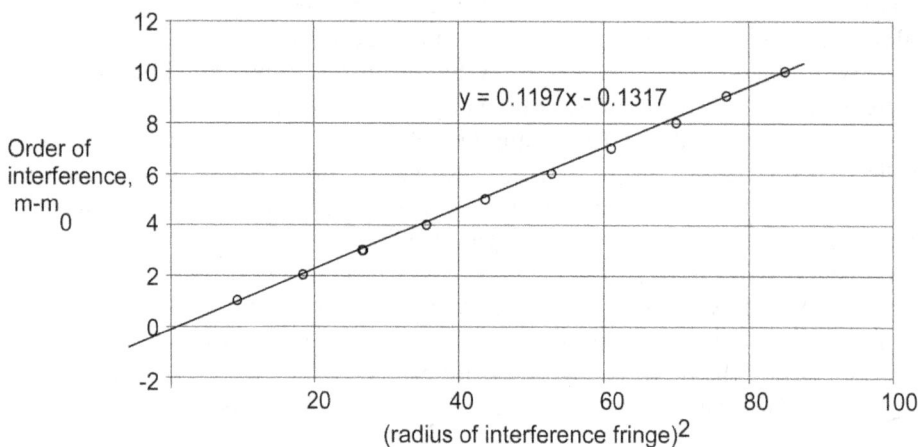

Figure 5.23. Plot of the fringe order m as function of $\cos \theta$, approximated by radius^2, from figure 5.19. The fractional part of the central order m_0 is seen to be -0.13 from this data.

various wavelengths chosen, using the thickness d as a variable. This way, the spacing can be found to an accuracy of about 0.1λ. The Benoit method can be used for a plate étalon (figure 5.18(a)) only if the refractive index of the plate is known accurately at each wavelength.

5.4.4 Applications

The Fabry–Perot can be applied to various laboratory experiments. The device is mainly used for observing extended spatially-incoherent sources; the fringes are localized at infinity. One application is to measuring the Zeeman effect. In a magnetic field of strength B, Zeeman splitting energies are typically $\delta E = \mu_B B$ where μ_B is the Bohr magneton, whose value is 9.3×10^{-24} J T. Compared with the photon energy at $\lambda = 600$ nm of order $E \approx 2$ eV $= 3.2 \times 10^{-19}$ J, we see that in a field of $B = 1$ Tesla, $E/\delta E = 3 \times 10^4$, indicating that the parameters discussed above (which gave $R = 10^6$) would be suitable for such an experiment. Small low-pressure Hg lamps which can be inserted between the poles of an electromagnet are available for this purpose. The dependence of the observed spectrum and its polarization on the direction of the emitted light with respect to the magnetic field are experiments which are important in understanding the quantum mechanics of the emitting atom.

A less demanding application is to measuring spectral fine structure, such as the separation of the lines in the Na doublet, where $E/\delta E \approx 1000$. In other cases, where the line to be investigated is weak or part of a coarser complex structure, the étalon is mounted in series with a monochromator which can be used to separate out a given spectral line for analysis. Sometimes, more simply, a narrow-band interference filter of the right wavelength can be used to separate the required spectral line.

One can also use a simple Fabry–Perot to measure the Doppler broadening of an emission line, of order $\delta E = k_B T$, where k_B is Boltzmann's constant. At room temperature, $E/\delta E = 70$, so that it is feasible to measure the broadening quite accurately as a function of temperature. One can easily observe and demonstrate the changes in the interference pattern from the 540 nm emission line of a mercury lamp as it warms up!

5.5 Holography with a digital camera

Traditionally, holography was carried out using high resolution photographic film, specially developed for the purpose with grains less than one micron in size. This film was not very sensitive, so that a relatively powerful laser was needed, and even so the film usually had to be exposed for many seconds. This long exposure required the optical setup to be very rigid and to be built on a vibration-isolated table, so that the relative movements of the components would be less than a fraction of a wavelength during the exposure time. Reconstruction of a visual image was done using the same laser impinging on the developed film hologram.

In recent years, the film has become hard to obtain, and holographic techniques have been developed which employ a digital camera rather than film. This development has both advantages and disadvantages [10].

On the positive side, a digital camera is much more sensitive than film, so that the exposure needed can be very short (less than 0.01 s) which obviates the need for an extremely stable and vibration-isolated optical system. Moreover, the electronic output of a digital camera is generally linearly related to the incident intensity over a greater range than in film[1]. Now the range of intensities recorded in a hologram lies between the sum of the object-beam and reference-beam intensities and the difference between them. Since digital processing of the hologram assumes this linear relationship, the ratio between object and reference intensities can be as large as 0.5, whereas a maximum of 0.1 was recommended for film. The image is reconstructed digitally (see appendix, section 5.5.6) and a variety of image processing algorithms can be used to improve it and to extract quantitative information from it. And the output is almost instantaneous; there is no film to be developed in the darkroom!

On the negative side, the resolution (pixel size) of a digital camera chip is usually of order 5–10 microns, which is at least an order of magnitude worse than the film,

---

[1] CCD sensors have a linear relationship at all intensities up to saturation; CMOS sensors are not quite as linear; see the examples in section 1.2.

so the number of pixels in a given area is then two orders of magnitude less than film. Moreover, the number of grey levels recorded by a pixel is more limited than in film. The first result of these limitations is that the angular field of view of an image is small. The camera has to record interference fringes whose period must be greater than two pixel-spacings (figure 5.24), which means that the reference beam and object beam can be separated by an angle at most $\lambda/(10 \ \mu m)$ radians, which is about 4°, compared with 40° or more for film; this limits the field of view considerably. Second, the reconstruction algorithm results in alias images which clutter the available image area; these aliases are the result of the periodic sampling of the hologram by the sensor structure, and do not occur with film because its structure is random (figure 5.25). Third, the angular resolution of the reconstructed image is

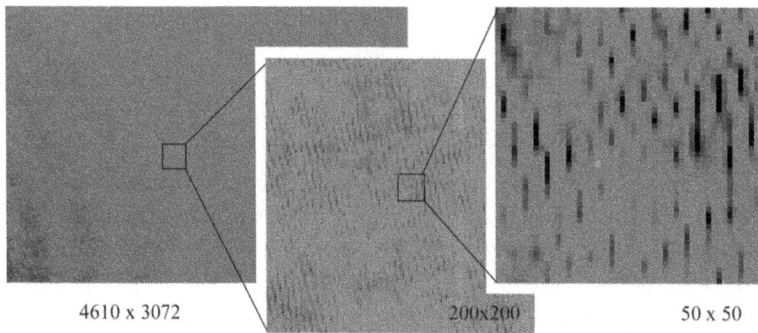

4610 x 3072   200x200   50 x 50

Figure 5.24. Hologram recorded on the digital camera. In the expanded displays one can discern interference fringes with a period of about 6 pixels.

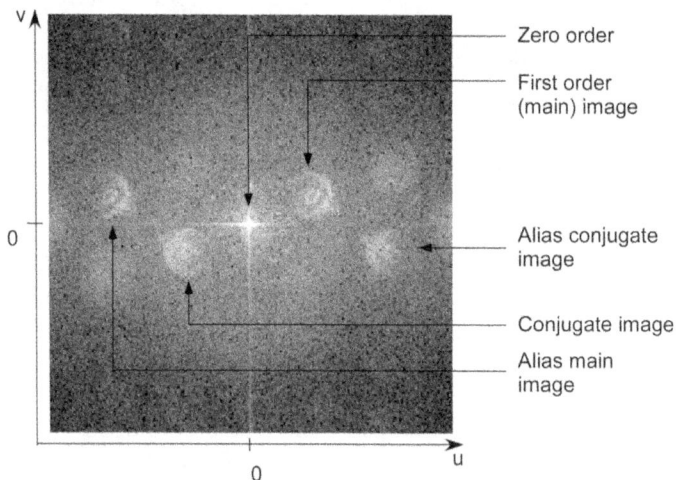

Figure 5.25. Reconstructions in the Fourier plane: The alias problem. This is the digital reconstruction from the hologram of a coin. The intensity scale is logarithmic, so that all the reconstructed beams can be visualized. Although there was only one coin in the field of view, it is reconstructed by the algorithm four times, and these reconstructions might overlap.

related to the size of the sensor in the same way as resolution by a telescope, $\lambda/$(linear dimension); this means that the number of resolvable pixels in the maximum reconstructed image area is equal to a quarter of the pixel number, typically 6 Mpx.

5.5.1 Experiments

An experimental investigation of the above features is a valuable way of under-standing image reconstruction. For a given object, we can do the following experiments:

1. Find the useful field of view, where there is no overlap between the basic image and the aliases or the zero order.
2. Determine the resolution of the image, and how it varies with the number of pixels used in the reconstruction of the hologram.
3. Show that the reconstruction is three-dimensional. One way to do this is to show that different parts of the image reconstruct sharply at their correct distances, and another way is to demonstrate parallax between the two reconstructions made using the left-hand half of the hologram and the right-hand half, respectively.

We will first discuss experiments using the off-line holographic set-up of Leith and Upatnieks [11]. We will then discuss in-line holography, the original Gabor [12] method, for which he received a Nobel prize in 1971.

5.5.2 Off-line (or side-band) holography

In this arrangement (figure 5.26), the expanded and spatially-filtered laser beam is split by a beam-splitter into a reference beam, which impinges directly on the camera, and a beam to illuminate the object which then scatters or reflects the light to the camera. The two mirrors in the system are arranged so that the path lengths from beam-splitter to camera by the two routes are roughly equal, so that maximum contrast fringes are obtained, although since the coherence length of the 5 mW He–Ne laser used is about the length of the laser, 30 cm, this equality is not critical. It is important that the beam-splitter be non-polarizing, because if the object and

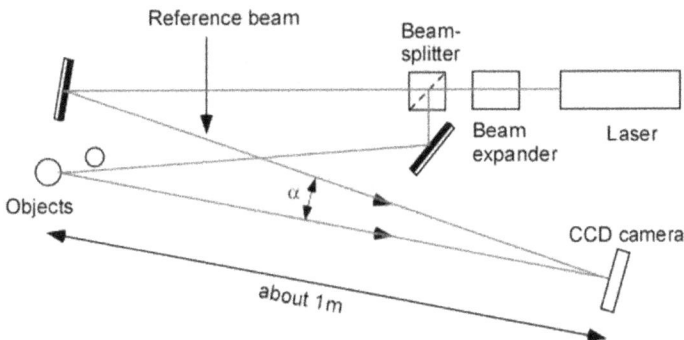

Figure 5.26. Optical layout for off-line holography.

reference beams have orthogonal polarizations they will not interfere. If there is a problem about polarization, a polarizer immediately in front of the camera will always solve the problem, at the price of about 60% of the light! A beam-splitter dividing the beam into two equal parts is appropriate if the object is quite reflective; however, if the object scatters diffusively, one really needs the beam illuminating the object to be much stronger than the reference beam. This can be achieved either by attenuating the reference beam or by using an unbalanced splitter—even a prism (4% reflection, 90% transmission) can be used for this, although one should be careful about the polarization of the reference beam if it is not reflected normal to the glass surface (figure 5.27). In film holography a ratio between the reference and object wave intensities of about 5–10 in the film plane is recommended so as to preserve linearity of the film recording, whereas the digital camera has linear response over its whole range (up to saturation); as a result, the ratio can be closer to unity in digital holography. This makes better use of the limited number of grey levels in the camera. If the camera has RGB colour response, one should note that there are twice as many green pixels as there are red ones, which gives some advantage to using the green channel rather than the red one (see figure 5.30).

The examples shown used coins as objects, since these reflect a considerable fraction of the light into the required direction. Moreover, they have a wealth of small detail which helps in focusing. Working with diffuse objects is a greater challenge. The exposure times used were of the order of 0.01 s, so that stability of the table posed no problem.

5.5.3 Reconstruction algorithm

Reconstruction of the images is carried out using a program in Matlab or another code. First, the hologram is read in (figure 5.24), and then one of the RGB colours is chosen for reconstruction. Although the He–Ne light is red, the hologram may contain over-exposed spots, and we have found that using the green channel often gives results which are better. A Cartesian (x,y) scale is then created (meshgrid) for the propagation calculation with the same size as the hologram matrix and with its origin at the centre. In the Fresnel approximation, and when the reference is a plane

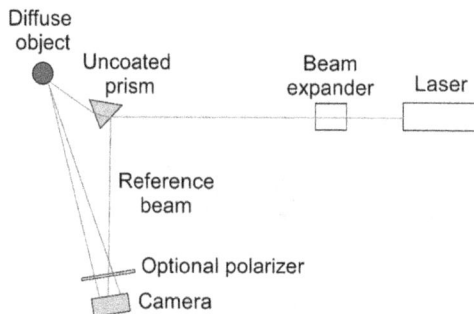

Figure 5.27. Optical layout for a diffuse object, using an uncoated prism as a beam-splitter (4% reflection, 90% transmission).

wave, the reconstructed image can be shown (see appendix in section 5.5.6) to be the Fourier transform of the product of the hologram and a quadratic phase factor $\exp[ik_0(x^2 + y^2)/2z]$ where $k_0 = 2\pi/\lambda$ and z is the distance from object to hologram. Different parts of the image of a three-dimensional object are 'in focus' when their z is correct. Since (x, y) is in pixel units, one should remember to use these units for k_0 and z as well. The absolute value of the reconstruction amplitude can first be displayed on a log scale (figure 5.25) to identify the main and conjugate reconstructions (first orders) and their aliases. Then one can concentrate on the region of the main image in order to improve it by adjusting z. This might not seem to be an adjustable parameter, but if the reference beam is not quite parallel or an incorrect value for the pixel size was used, the errors can be eliminated by adjusting z until the best contrast is obtained.

5.5.4 Experimental aims

Some suggested directions for experiments are

a. Record a hologram and write a program to reconstruct the image of a reflecting object, using the Fresnel approximation. Then show that the resolution of the image depends on the size of the hologram, by reconstructing images from progressively smaller regions of the hologram (figure 5.28). Also estimate the depth of focus by varying the value of z used in the algorithm.

b. Make a three-dimensional object, e.g. two coins at different distances, and illustrate the three-dimensional aspects of the reconstruction. Parallax can be observed between reconstructions made from the left and right halves of the hologram. (figure 5.29)

c. Experimental determination of the field of view.

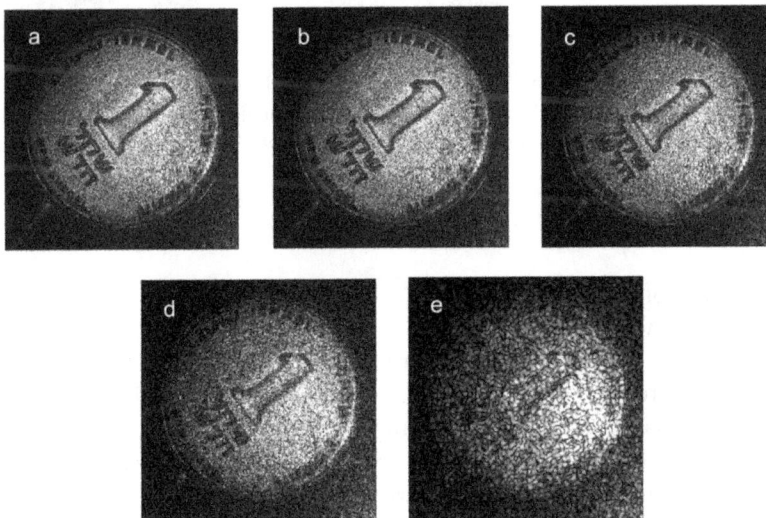

Figure 5.28. Showing how the resolution of the image depends on the size of the region of the hologram used for the reconstruction. (a) Reconstruction from the complete 4610×3800 pixel hologram; (b) 2000×2000 pixels; (c) 1500×1500 pixels; (d) 1000×1000 pixels; (e) 500×500 pixels.

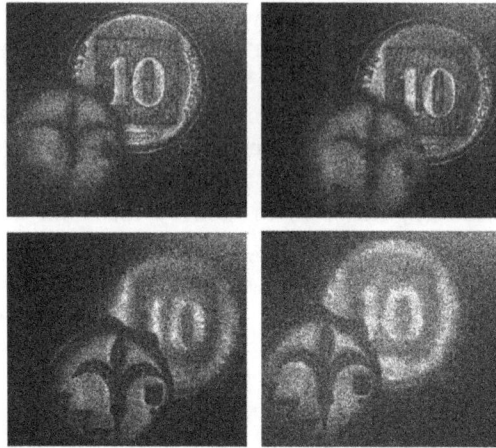

Figure 5.29. Depth of focus and parallax. Reconstructions of images of two coins at different distances from a single hologram. The left images are from the left half of the hologram and the right images are from the right half. The upper pair use a distance z in the reconstruction algorithm appropriate to the more distant coin, and the lower pair use a factor appropriate to the closer coin. Notice both the sharpness of the images and the parallax when moving from left to right.

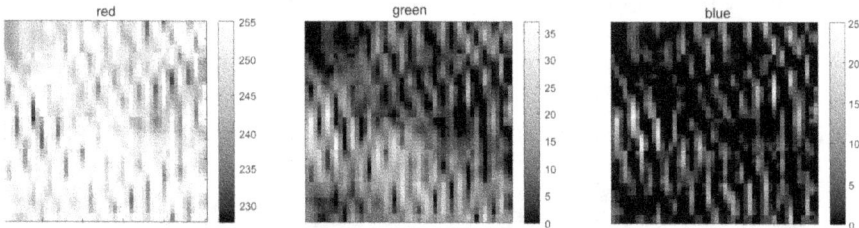

Figure 5.30. Using the red, green or blue channel: images of a small part of the hologram recorded with red laser light. Notice that the red channel has very low fringe contrast or visibility $V = (I_{max} - I_{min})/(I_{max} + I_{min})$ = 25/487 = 0.05, whereas the green has much better (1.0), and the blue also has good contrast, but the signal is weaker.

    d. Removal of artefacts such as the zero order and aliases by making two holograms with the same set-up, with a phase difference of π between their reference beams. The phase difference can be introduced easily by including in the reference beam an optical window which can be rotated about a vertical axis by an appropriate angle. Then, the image is reconstructed by using the difference between the two holograms. If the reference and object beam amplitudes are A and a, respectively, One hologram records $|A + a|^2$ and the other records $|A - a|^2$, and the difference is then $4Aa$ which can be shown to eliminate some artefacts.

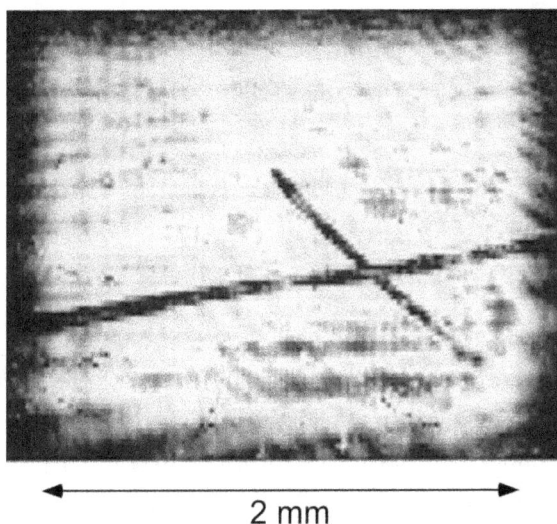

2 mm

Figure 5.31. In-line holographic reconstruction of the image of a pair of stretched hairs. The object was 5cm from the pinhole and the camera was at 50cm.

e. Make and reconstruct a hologram from a diffusely-scattering object, such as a small toy or chess piece (figure 5.27). Getting a good reconstruction is considerably more difficult than it was for highly-reflecting coins.

5.5.5 In-line holography

The setup for in-line holography is simpler than that for off-line holography, but the choice of objects to reconstruct is more limited. Essentially the object has to be transparent, with a large ratio between the background and object parts. For example, one can stretch some thin wires across a frame (figure 5.31) or use a tank filled with water with some scattering particles as three-dimensional objects. Reconstruction of the sharp image is similar to that for the off-axis holography, but in this case the main image, conjugate image and zero order overlap. However, if the hologram plane is not too far from the object, the latter two are sufficiently smeared that the overlap is not too much of a problem. See the article by Garcia-Sucerquia *et al* [13] for a discussion; these authors also improved on the Fresnel approximation in their work.

5.5.6 Appendix. Derivation of the reconstruction procedure in the Fresnel (small angle) approximation

The object is $f(x)$ in one dimension in the plane $z = 0$. Light propagates to the plane z, where the hologram is recorded. The Fresnel pattern in that plane is $h(x')$. Using a spherical wave propagating from each point on the object we have

$$h(x') = \exp[ik_0 z] \int f(x) \exp\left[\frac{ik_0(x'-x)^2}{2z}\right] dx$$

$$= \exp[ik_0 z] \int f(x) \exp\left[\frac{ik_0(x'^2 - 2xx' + x^2)}{2z}\right] dx$$

$$= \exp\left[\frac{ik_0 x'^2}{2z} + ik_0 z\right] \int f(x) \exp\left[\frac{ik_0 x^2}{2z}\right] \exp\left[\frac{-ik_0 xx'}{z}\right] dx$$

$$= \exp\left[\frac{ik_0 x'^2}{2z} + ik_0 z\right] . \ F. \ T. \ \left\{ f(x) \exp\left(\frac{ik_0 x^2}{2z}\right) \right\}$$

where the variable of the Fourier transform (F.T.{ }) is $k_0 x'/z$ which is $u \equiv k_0 \sin \theta$. By means of the holographic technique of interference with the reference wave A, we record in the intensity the products $A^*h(x')$ and $Ah^*(x')$ including their phases. As a result, we can now take the inverse Fourier transform (I.F.T.{ }) of $h(x') \exp\frac{ik_0 x'^2}{2z}$ to get the reconstructed image $f_r(x)$. In off-line holography, the reconstructions of $A^*h(x')$ and $Ah^*(x')$ are spatially separated, and in in-line holography they are longitudinally separated:

$$f_r(x) \exp\left(\frac{ik_0 x^2}{2z}\right) = \ I. \ F. \ T. \ \left\{ h(x') \exp\frac{ik_0 x'^2}{2z} \right\}.$$

(Note that we have used 'I.F.T.' here, which is the inverse Fourier transform. It is the same as 'F.T.', except that in the latter, the axes are inverted, i.e. $x' = -x$, $y' = -y$.) Finally, the intensity of the reconstructed image is

$$|f_r(x)|^2 = \left| \ I. \ F. \ T. \ \left\{ h(x') \exp\frac{ik_0 x'^2}{2z} \right\} \right|^2.$$

In words, we take the hologram as it is photographed in the x' plane, multiply it by the spatial function $\exp\frac{ik_0 x'^2}{2z}$, and take its inverse Fourier transform. This is the picture shown in figure 5.25 (for the same process in two dimension, $f(x,y)$). It is important to remember in writing an algorithm for this calculation that all variables, including k_0 and z, must be translated to pixel units. When the picture (figure 5.25) is displayed, one plays around with the value of z to get the sharpest image. Note that the conjugate image is also produced, which is blurred because in the quadratic phase factor $\exp\frac{-ik_0 x'^2}{2z}$ in $Ah^*(x')$ has the wrong sign.

If we want to use this for interferometric holography (section 5.6), for example, where the phase of the reconstruction is important, we would write

$$f_r(x) = \exp\left(\frac{-ik_0 x^2}{2z}\right) I. \ F. \ T. \ \left\{ h(x') \exp\frac{ik_0 x'^2}{2z} \right\}.$$

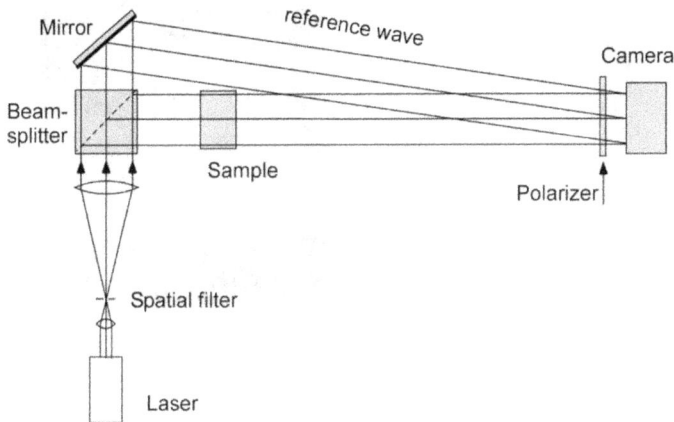

Figure 5.32. Setup used for holographic observation of strain patterns in a sample. Remember that the angle between the object and reference beams has to be small, as in all digital holography.

5.6 Interferometric holography

Holography provides us with a very important technique for interferometry. Since a hologram preserves information about the phase of a wavefront (relative to the reference wave) interference between two wavefronts can be observed by interfering each of them sequentially with a given wavefront, and then reconstructing the difference between the two holograms simultaneously [10]. The observed reconstruction is then the difference between the two complex-valued wavefronts, which is their interference pattern.

5.6.1 Double exposure holographic interferometry

A very simple experiment which can be carried out this way involves the photoelastic effect (section 7.6). A sample made from a transmitting plastic plate is used as the object in a holographic setup (figure 5.32). The sample is mounted in a frame which allows it to be strained (figure 5.33(a)).

First, a hologram is recorded in the unstrained situation. Then a second hologram is recorded when the sample is strained by the frame. When the difference[2] between the holograms is reconstructed, the phase changes resulting from the change in refractive index due to the photo-elastic effect, which are proportional to the local strain, become immediately evident. Since the photoelastic effect is dependent on the polarization of the incident light, the experiment works best if a polarizer is placed before the camera recording the hologram. Actually, it turns out that the sensitivity of this method to a given stress in Perspex is about 20 times higher than that of the conventional photoelastic effect using a polarizer–analyser pair. The same idea can be used to observe the dimensional changes of an object when its temperature is changed (e.g. by using a transparent light bulb, which is turned on between the two

---

[2] Either the sum or difference between the holograms can be used, but the difference has the advantage that the zero order in the reconstruction is minimized, so that the field of view is largest.

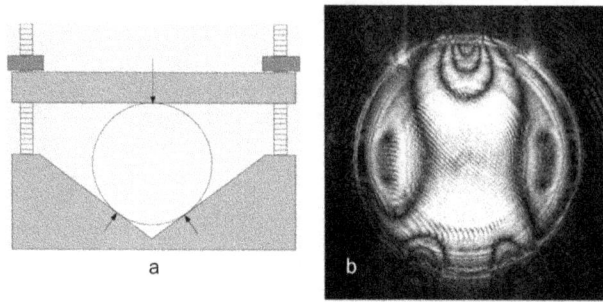

Figure 5.33. (a) The frame used to introduce stress in a 20 mm diameter Perspex disc; (b) induced strain pattern observed on application of a small stress by finger-tightening the nuts (in red).

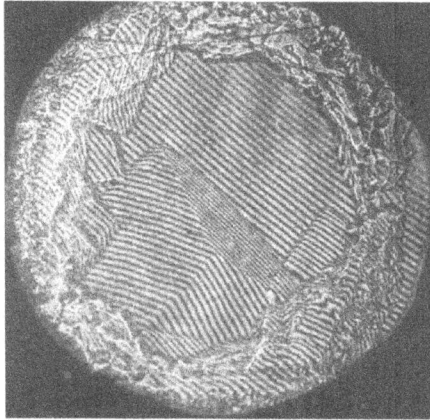

Figure 5.34. Holographic interferograms of helium crystals growing in a transparent cell from superfluid, at temperature 0.55 K and pressure 25 bar. A glass plate hologram of the liquid-filled cell was made at a lower pressure, below the melting pressure, and then the hologram was returned to its original position in the optical system after developing, and the pressure was raised to form the crystals. The image is the interference between the reconstruction of the liquid-filled cell and the image of the cell containing the crystals [14]. Reproduced from [9] with permission of Cambridge University Press © 2011.

exposures), or phase changes in transparent media, etc. The great advantage of holographic interferometry for these experiments is that features of the system which do not change between the exposures (such as the flatness of the surfaces of the Perspex plates used) do not appear in the interference pattern. An example of the method [14] (using photographic plate holograms, in 1980) was to observe the formation of solid helium crystals when the pressure was increased in a cell containing superfluid liquid helium (figure 5.34).

5.6.2 Time exposure holography

Another way in which holographic interferometry has been widely used is to observe normal mode patterns in vibrating objects. Light is reflected from the surface of a plate, for example, which can be induced to vibrate transversely. The light intensity used is sufficiently weak that the camera exposure needed to record its hologram is several times longer than the period of vibration induced in the plate. As a result, when the holographic image is reconstructed, regions where the vibration amplitude is greater than half a wavelength of the light will have essentially zero intensity, and the bright parts of the reconstruction correspond to the nodes of the vibration pattern, where the surface does not move significantly. A series of reconstructions made for various resonant frequencies can be compared to calculated normal modes of vibration.

5.6.3 A comment on holographic interferometry from the point of view of wave–particle duality

A simple experiment to carry out involves making the holograms of a pair of coherently illuminated slits. In the apparatus of figure 5.32, we put a sample consisting of two narrow parallel slits. Then we can make a hologram of the pair of slits, H_{pair}, and when we reconstruct the field at infinite distance we get the Young's interference pattern between them. Now we make individual holograms of the two slits, H_{left} and H_{right} when the other slit is blocked. The reconstruction of either of these is a uniform light field, but if we reconstruct $H_{left} \pm H_{right}$ we get the Young's fringes again. Of course, from the point of view of wave theory the interpretation is obvious. But if the holograms H_{left} and H_{right} were real transparencies, and we placed them in contact for the joint reconstruction, how would you interpret this from the point of view of photons?

5.7 Computer-generated holography

There are many examples of 'virtual objects' which can be visualized in three dimensions by calculating the hologram of the object, and then reconstructing it in the usual manner. The reconstruction requires the calculated hologram to be printed on film, for example, or displayed on a spatial light modulator (SLM) which modulates an incident reference wave in accordance with the holographic data. Two important examples of applications are:

1. In medicine, information gained from a CT examination of a patient, which provides a series of two-dimensional slices of the image of the body, is used to create mathematically a hologram of the patient's internal structure. The hologram is then reconstructed using an SLM as a three-dimensional image which can be viewed from any direction. The reconstruction can be used by surgeons to plan an operation in detail, for example.
2. Optical testing of an arbitrary surface (an aspheric lens, for example) can be done by comparing the manufactured surface with a hologram of the required surface, using holographic interferometry. The hologram is made by calculation, using the required surface formula and a reference wave as input.

The hologram is constructed in a computer program in the following way. First, the object is chosen—let's say, the letter **H**. This can be described as a function $H(x, y, 0)$ in the $z = 0$ plane, where $H = 1$ on the lines of the letter, and $H = 0$ in the spaces surrounding it. This function is then considered as a slide in the $z = 0$ plane, and is 'illuminated' by a plane wave, so that the wave received in a plane where $z \neq 0$ is its Fresnel diffraction pattern:

$$E(x, y, z) = \exp(ik_0 z)$$

$$\times \iint H(x', y', 0) \exp\{-ik_0[(x - x')^2 + (y - y')^2]/2z\} dx' dy'.$$

We now add to this the reference wave, which we model as a plane wave of amplitude E_0, propagating in the x–z plane at an angle α to the z-axis, $E_R(x, y, z) = E_0 \exp[ik_0(z \cos \alpha + x \sin \alpha)]$, where E_0 is about 5 times larger than the maximum amplitude of $E(x, y, z)$ in the plane z, in order to create approximate linearity when the hologram value is squared in the next stage. It is possible to create an in-line hologram by putting $\alpha = 0$, but the experiment is more instructive with non-zero α. The hologram amplitude is thus $E_H = E_R + E$. Finally, we calculate the light intensity $I(x, y, z) = |E_H|^2$ in the z-plane, and print it as a figure. The figure is now photographed as a slide with size like that of the laser beam to be used for the reconstruction; this is the hologram. Alternatively, the function is fed into an SLM. It is worth looking at the hologram under a microscope. You should be able to see the fringes which contain the information, and have a period $\lambda/\sin \alpha$.

5.7.1 Reconstruction

When the slide or SLM is illuminated by a plane wave along the z-axis, three distinct waves will be created. One is a wave along the z-axis, which is the zero order of diffraction. Then there will be a first order wave, at angle α to the z-axis, and in the plane z it will focus to a real image of the original E function. There will also be a $-$first order wave, at angle $-\alpha$, which can be seen to originate from a virtual image in the plane $-z$. If you photograph the two reconstructions, which one is closer to the original? It is possible to repeat this experiment using a reference beam which is not a plane wave, for example a wave diverging from a point. When the reconstruction is done with the real-life implementation of this wave, the virtual image will come out right, but the real image may be distorted. This suggests the idea of using holography as a method of coding in which the hologram of a picture is made using a known complicated reference wave. The picture can then be reconstructed only if the reference wave (the key) is known.

5.7.2 Three-dimensional object

Clearly, there is no problem in repeating this exercise for an object H which is a function of three dimensions. We then get:

$E(x, y, z) =$

$$E_0 \iiint \exp[ik_0(z - z')]H(x', y', z') \exp\{-ik_0[(x - x')^2 + (y - y')^2]/2(z - z')\} dx' dy' dz'.$$

Then a detailed picture of any particular plane in the image can be reconstructed using an appropriate reconstruction wave. Note that the other planes in the image will simultaneously be reconstructed, but they will not be in focus.

References

[1] Michelson A A 1927 *Studies in Optics* (New York: University of Chicago Press) reprinted (1995) by Dover Publications
[2] Hariharan P 2007 *Basics of Interferometry* 2nd edn (New York: Academic)
[3] Armitage J D and Lohmann W 1965 Rotary shearing interferometry *Opt. Acta* **12** 185–92
[4] Schwartz E 2017 *Eur. J. Phys.* **38** 015301
[5] Malik Z, Cabib D, Buckwald R A, Talmi A, Garini Y and Lipson S G 1996 Fourier transform multipixel spectroscopy for quantitative cytology *J. Microsc.* **182** 133
[6] Schröck E *et al* 1996 Multicolor spectral Karyotyping of human chromosomes *Science* **273** 494–7
[7] Hernandez G 1986 *Fabry–Perot Interferometers Cambridge Studies in Modern Optics* (Cambridge: Cambridge University Press)
[8] Tolansky S 1973 *An Introduction to Interferometry* (New York: Wiley)
[9] Lipson A, Lipson S G and Lipson H 2011 *Optical Physics* 4th edn (Cambridge: Cambridge University Press)
[10] Schnars U and Jueptner W 2005 *Digital Holography* (Berlin: Springer)
[11] Leith E N and Upatnieks J 1962 Reconstructed wavefronts and communication theory *J. Opt. Soc. Am.* **52** 1123–30
[12] Gabor D 1948 A new microscopic principle *Nature* **161** 777–8
    Gabor D 1949 Microscopy by reconstructed wavefronts *Proc. R. Soc.* **197** 454–87
[13] Garcia-Sucerquia J, Xu W, Jericho S K, Klages P, Jericho M H and Jürgen Kreuze H 2006 *Appl. Opt.* **45** 836
[14] Landau J, Lipson S G, Määttänen L M, Balfour L S and Edwards D O 1980 Interface between Superfluid and Solid ^{4}He *Phys. Rev. Lett.* **45** 31

IOP Publishing

Optics Experiments and Demonstrations for Student Laboratories

Stephen G Lipson

Chapter 6

Physical optics III: topics in wave propagation

6.1 Optical tunnelling: frustrated total internal reflection

Optical tunnelling is the optical analogue of quantum mechanical tunnelling by which a particle incident on a potential barrier higher than its kinetic energy, but which has finite thickness, has a non-zero probability of being found on the other side of the barrier. In the optical scenario, we have an interface between two media of different refractive indices, such as glass and air. The light is incident from the side of the higher index, glass, at an angle above the critical angle, so that total internal reflection occurs. However, in the second medium there is a decaying evanescent wave induced, which transports no energy if it propagates to infinity. The presence of a second interface with a higher-index medium within a short distance of the first, where the evanescent wave amplitude is still appreciable, results in creation of a travelling wave which transports energy through the gap between the media (figure 6.1). This is called 'frustrated total internal reflection' or 'optical tunnelling'. The phenomenon was apparently known to Newton and was studied long before quantum mechanics was discovered [1, 2], but often students do not meet the optical topic until they have learnt about it in a quantum environment. When the matter waves of quantum mechanics are replaced by electromagnetic waves, a similar process is possible when the potential barrier is replaced by a function of the refractive index and the angle of propagation. In quantum language, we find that photons travelling in a given medium, which should have been totally reflected by internal reflection at an interface with a thin slab of a medium of lower refractive index, have a non-zero probability of penetrating it, provided that the second medium has a limited thickness.

doi:10.1088/978-0-7503-2300-0ch6

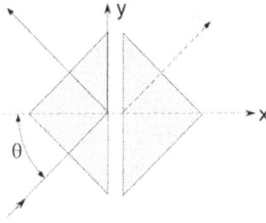

Figure 6.1. Tunnelling experiment using two prisms.

6.1.1 Theory of optical tunnelling

A simple way to see the correspondence is to consider a plane electromagnetic wave travelling at angle θ to the x-axis in a medium in which the refractive index $n(x)$ is a function of x. Then, from Snell's law, $\theta(x)$ is also a function of x. The electromagnetic wave equation in two dimensions (x, y) is

$$\frac{\partial^2 E}{\partial x^2} + \frac{\partial^2 E}{\partial y^2} = \frac{n^2(x)}{c^2}\frac{\partial^2 E}{\partial t^2}.$$

Since in planes of constant x the wave has the form $E = E_0 \exp[i(k_y y - \omega t)]$ where $\frac{\omega^2}{c^2} = k_0^2$ and $k_y = k_0 n(x)\sin \theta(x) = k_0 n_0 \sin \theta_0$ from Snell's law, n_0 and θ_0 representing the incident medium, we can write the above equation as

$$\frac{\partial^2 E}{\partial x^2} - k_0^2 n^2(x)\sin^2 \theta(x)E = \frac{\partial^2 E}{\partial x^2} - k_0^2 n_0^2 \sin^2 \theta_0 E = -n^2(x)k_0^2 E.$$

This can be reorganized as an equation which is analogous to Schrödinger's equation:

$$\frac{\partial^2 E}{\partial x^2} + \left[-k_0^2 n_0^2 \sin^2\theta_0 + n^2(x)k_0^2\right]E = 0 \quad \text{Maxwell}$$

$$\frac{\hbar^2}{2m}\frac{\partial^2 \psi}{\partial x^2} + [E_t - V(x)]\psi = 0 \quad \text{Schrödinger.}$$

We see that the total energy of the particle E_t in Schrödinger's equation is replaced by $-k_0^2 n_0^2 \sin^2 \theta_0$, which is invariant from medium to medium. The potential $V(x)$ is replaced by $-k_0^2 n^2(x)$, which means that a lower refractive index n can simulate a potential barrier in a higher refractive index region. To simulate tunnelling, we then require that the incident conditions have negative $k_0^2 \sin^2 \theta_0$, which means that incidence is in a medium above the critical angle at an air interface. Such conditions can be obtained by using a prism, as in figure 6.1, which shows the usual experiment for investigating tunnelling using microwaves. The wave in the air gap therefore has an exponential, rather than a sinusoidal, dependence on the propagation distance.

The value of k within the air gap is calculated as follows. For the situation shown in the figure, where $\theta = 45°$, we have $k_y = k_0 n \sin 45° = k_0 \sin \theta_0$ in the air gap. There, $k_x = k_0 \cos \theta_0 = k_0\sqrt{1 - \sin^2 \theta_0} = \pm i k_0\sqrt{n^2 \sin^2 45° - 1}$. If $n > \sqrt{2}$, k_x is imaginary.

As a result, the wave in the air gap decays (or is amplified) with a decay distance $l = 2\pi/[k_0\sqrt{n^2/2 - 1}] = \lambda/\sqrt{0.125} \approx 2.8\lambda$ (for $n = 1.5$).

6.1.2 Visualizing tunnelling in a Newton's rings configuration

In order to visualize how the tunnelling intensity varies with distance, the experiment can be done using a curved second surface, as in the Newton's rings experiment, so that the separation of the two surfaces is a function of position (figure 6.2). The setup shown in figure 6.3, in which the reflected light (i.e. that does *not* tunnel) is observed is most convenient, and the dependence on the exact angle of incidence is instructive. Note that, as in all thin film interference, the interference phenomenon is localized in the region of the contact and therefore the camera must be focused on this region by means of an imaging lens.

Figure 6.4 shows two illustrations of observations; in the first, the incidence of the plane wave on the prism hypotenuse is at $\theta < \theta_c$ and in the second at $\theta > \theta_c$. Knowing the radius of curvature of the lens surface, which can be deduced from the Newton's rings, the tunnelling decay distance l can be measured.

In order to make a quantitative measurement of the tunnelling distance, in terms of the wavelength, it is necessary to take two photographs: first, the image of the

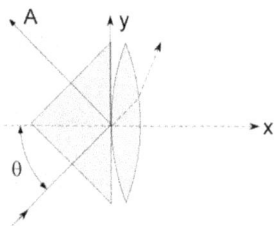

Figure 6.2. Varying the tunnelling distance spatially.

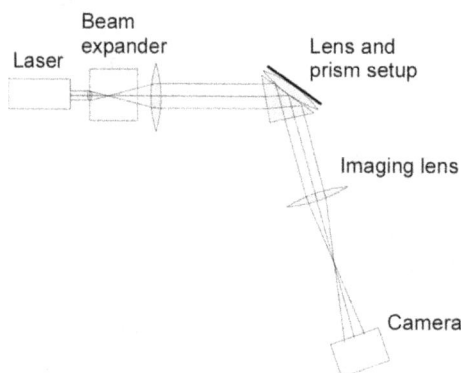

Figure 6.3. Setup to observe tunnelling at incidence close to the critical angle.

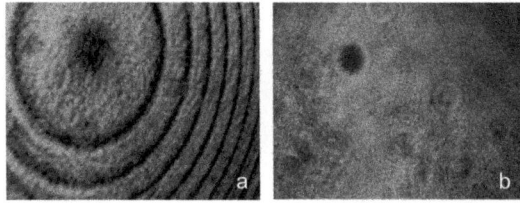

Figure 6.4. (a) $\theta < \theta_c$ and (b) tunnelling when $\theta > \theta_c$. Compare to figure 6.6.

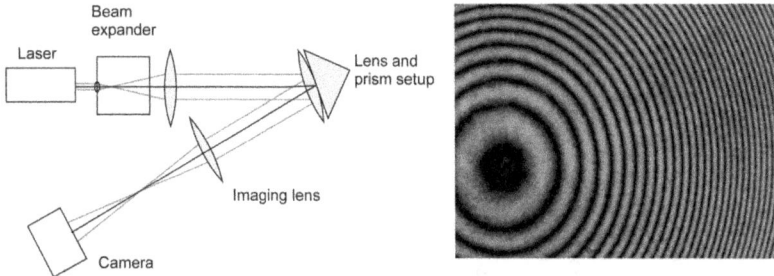

Figure 6.5. (a) Setup for observing Newton's rings and (b) Newton's rings at almost normal incidence.

tunnelling region as in figure 6.4; then, using the same imaging optics, a photograph of the Newton's rings for normal incidence to determine the thickness of the tunnelling gap as a function of position. The latter rings are much tighter than the Newton's rings close to the critical angle, since the projected wavelength in the normal direction is $2\pi/k_0 = \lambda$ at normal incidence, but $2\pi/(k_0 \cos \theta_0) = \lambda/\cos \theta_0$ near the critical angle, diverges as $\cos \theta_0 \to 0$. Photographing the rings at normal incidence with a laser is not trivial, because the rings are weak compared with other parasitic reflections (see the blue curves in figure 6.6)! The easiest way is to observe the Newton's rings from the opposite side of the prism-lens assembly at a small angle of incidence (figure 6.5). Alternatively, the rings can be photographed using the same imaging system when the lens surface is contacted with a plane glass window instead of the prism; the disadvantage of this is that the lens-prism system has to be disassembled. In this case, the parasitic fringes can be avoided if the window surfaces have a small angle between them.

6.1.3 Interpreting the results

At first sight, it might seem that for incidence at the critical angle the transmitted wave function has infinite decay depth and therefore the tunnelling spot should be very broad. However, at the same angle the reflection coefficients at the air glass interfaces become unity, and so the fringes observed are not from two-beam interference but from multiple-beam interference. As a result, the fringes are Fabry–Perot fringes, which are much narrower than the sinusoidal fringes of two-beam interference. You can see this on comparing the fringe profiles in figures 6.4(a) and 6.5(b). The complete theory is given in the paper by Vörös and Johnsen [3]. It

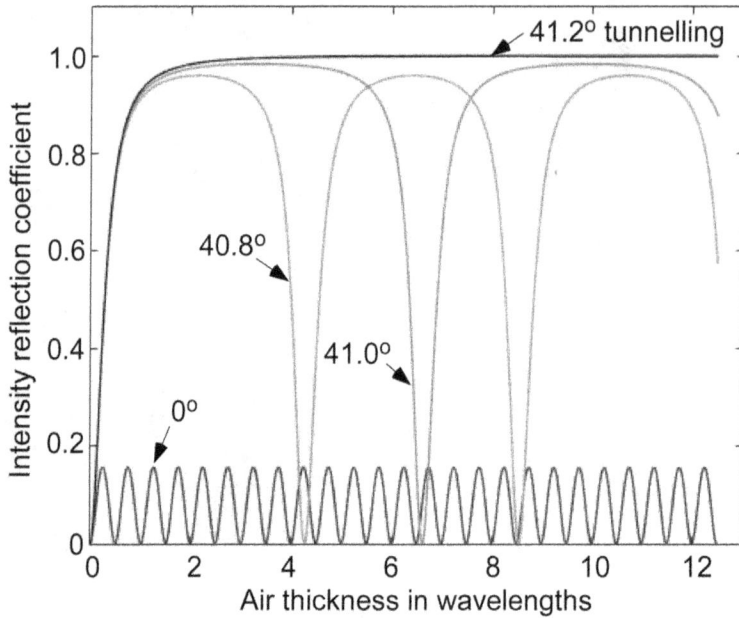

Figure 6.6. Reflected intensity profiles as a function of tunnelling thickness in air between two glass ($n = 1.52$) surfaces at incidence angles 0°, 40.8°, 41.0° and 41.2°. The critical angle is 41.13°.

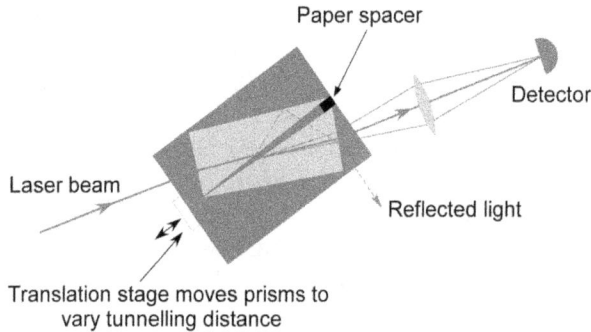

Figure 6.7. Using a pair of prisms to investigate optical tunnelling.

turns out that, essentially, there is a continuous transition from the central dark spot of the Newton's rings (figure 6.4, left) and the tunnelling spot (right) as one goes through the critical angle. Figure 6.6 shows numerical calculations of the tunnelling intensity as function of spacing between the surfaces, which confirm this continuous transition.

6.1.4 Direct measurement of the tunnelling probability

A quantitative investigation of the tunnelling phenomenon, as described by Vörös and Johnsen [3], uses an air wedge formed between two prism surfaces, as shown in figure 6.7. The wedge has a thickness which varies linearly with position (as opposed to the quadratic variation in the lens–prism contact) and is investigated by a narrow laser beam as the system is translated relative to it. It is important to avoid the detector's seeing the reflected light, which is dominant; the addition of a collector lens helps to do this.

6.2 The acousto-optic effect

The acousto-optic effect couples sound waves, or phonons, with photons. In terms of classical concepts, a sinusoidal sound wave of frequency f creates a periodic pressure variation with wavelength $\Lambda = v_s/f$ which travels through the material at the sound velocity v_s. This pressure variation results in changes in density and therefore in refractive index, so the material behaves as a three-dimensional phase diffraction grating. The grating diffracts an incident monochromatic light wave into a series of diffraction orders, and it is the intention of the experiment to investigate this process quantitatively. Since the sound wave is dynamic, there is also a Doppler effect associated with the diffraction, which results in a shift in frequency which is different for each order, but this is a very small effect which will be ignored in the experiments. Acousto-optic cells are often used for deflecting light beams, because the frequency of the sound wave is easily varied and therefore the angles of diffraction can be controlled. In addition, since the grating is a phase grating, diffraction efficiencies approaching 100% can be achieved. Devices employing the acousto-optic effect use crystals such as lead germanate and lithium niobate, but the effect is also quite large in water, which is convenient for laboratory experiments and it is also isotropic.

The general problem of diffraction by a three-dimensional grating is hard to solve rigorously, but there are basically two regimes of diffraction, which apply to different experimental parameters. These are called the Bragg regime (in which the three-dimensional structure is taken into account) and the Raman–Nath regime, which is a two-dimensional approximation. The full theory is discussed by Born and Wolf [4, chapter XII].

In the Bragg regime, which generally applies to acoustic frequencies above about 100 MHz, the incident light is reflected specularly by the acoustic wavefront planes when Bragg's law is satisfied, i.e. when the incident angle satisfies $m\lambda = 2n\Lambda \sin \theta$. In water, where $v_s = 1500$ m s^{-1}, this would be satisfied by $\theta \approx 2m \times 10^{-2}$ rad for a 100 MHz acoustic wave and light with wavelength 600 nm. Since the acoustic wave gives rise to a sinusoidal phase grating, the amplitude of the wave can in principle be adjusted till the reflection is almost complete. The Bragg regime is not considered in the experiment described below.

6.2.1 Experiments in the Raman–Nath regime

We use a simple geometry, consisting of a rectangular metal cell with two parallel windows, to which a piezo-electric (PZT) disc transducer is attached on one face normal to the windows (figure 6.8). The cell is filled with distilled water. When the transducer is activated by a signal generator, a plane acoustic wave propagates

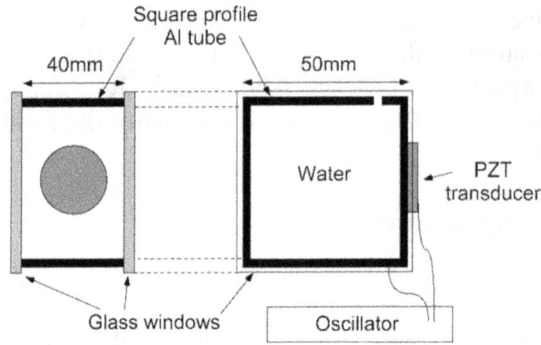

Figure 6.8. Side and front views of the experimental cell.

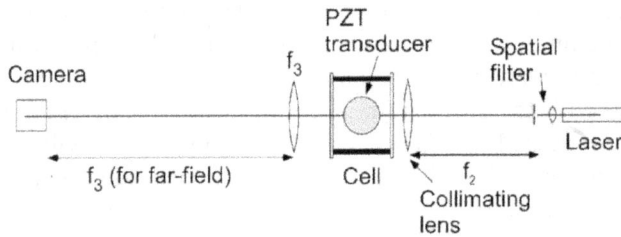

Figure 6.9. Diffraction optics set-up.

across the cell and is partially reflected at the far side. The wave has wavefronts separated by the acoustic wavelength $\Lambda = v_s/f$; for example, at frequency $f = 1$ MHz in water, $\Lambda = 1.5$ mm. At certain frequencies, standing waves are created. The cell is investigated using conventional diffraction optics with a He–Ne laser, and both Fraunhofer (far-field) and Fresnel (near-field) diffraction patterns can be recorded (figure 6.9). Knowing the velocity of sound in water, the frequency of the acoustic wave and the camera chip dimension and pixel size, it is easy to choose an appropriate focal length f_3 for the diffraction lens.

In the Raman–Nath approximation, for this cell we consider diffraction by a phase grating which is sinusoidal along the axis x normal to the light propagation. The grating has period Λ and phase amplitude A, where A is proportional to the pressure wave amplitude. The grating is the result of multiple reflections of the sinusoidal longitudinal acoustic waves at two parallel walls of the cell. The reflection coefficient R for the wave at the cell walls at $x = \pm L$ is quite close to -1, and the wave displacement which is excited after multiple reflections has the form

$$\xi = \xi_0\{\exp(iqx) + R\exp[iq(2L - x)] + R^2\exp[iq(4L + x)]$$
$$+ R^3\exp[iq(6L - x)] + \ldots\}$$

where $\Lambda = 2\pi/q$. The local pressure p is proportional to $d\xi/dx$. When $R \approx -1$ this sums to

$$\xi = 2\exp(iqL)\xi_0\sin[q(x - L)]/[1 - R^2\exp(4iqL)].$$

This form has wavelength Λ and its amplitude is resonant, peaking when $4qL = 2m\pi$, i.e. when $\Lambda = 4\pi/q = 4L/m$, implying that the width of the cell $2L$ is an integer multiple of half-wavelengths (a standing wave, with zeros at $x = -L$ and L). The resonance peaks are quite sharp. The optical phase amplitude A is proportional to p, where the constant of proportionality S involves the compressibility of the water, the length of the cell along the optical axis and the relationship between refractive index and density of the water, and so the index amplitude has maximum value at the walls of the cell.

The grating function is thus $f(x) = \exp[iA \sin(2\pi(x - L)/\Lambda)]$ where $A = Sp_0$. In the far field, the diffraction pattern is the Fourier transform $F(u) = \sum_m \delta(u - m2\pi/\Lambda)J_m(A)$

where J_m is the mth order Bessel function and for normal incidence $u = k_0 \sin \theta$. This diffraction pattern consists of the set of orders with intensities $|J_m(A)|^2$. In particular, notice that $J_0(A) = 0$ when $A = 2.4, 5.5, 8.7$, etc, which means that amplitudes might exist where the whole of the wave is diffracted and there is no zero order, although such amplitudes may be inaccessible because of experimental limitations. Since R is close to -1, prominent standing waves are excited and these are most convenient for experimentation. The period Λ is equal to the wavelength of the acoustic wave, i.e. 1.5 mm in the above example. In addition, the wave is reflected and scattered from the other boundaries of the cell, and an acoustic wave pattern exists which is more complicated than a simple plane wave, but is made up of superpositions of waves with the same wavelength, but angle-dependent intensity. This results in additional orders in the Fourier plane which lie on circles around the main orders when the amplitude is large enough (figure 6.10).

When the sensor does not lie in the Fourier plane (focal plane of the diffraction lens) we get a Fresnel or near-field diffraction pattern, which contains caustics resulting from the phase distribution in the emerging wave. These show the acoustic wave amplitude clearly. Two examples are shown in figure 6.11.

As pointed out by Ribak [5], these caustics are very sensitive to aberrations in the optical system (lenses and cell windows) and can be used to quantify them.

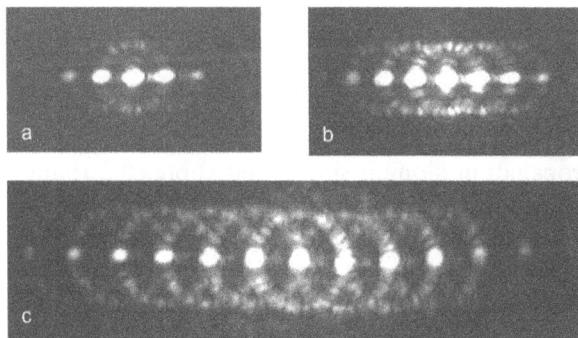

Figure 6.10. Far-field diffraction patterns: (a) off-resonance, (b) on-resonance around 1.66 MHz and (c) on resonance at 2.33 MHz.

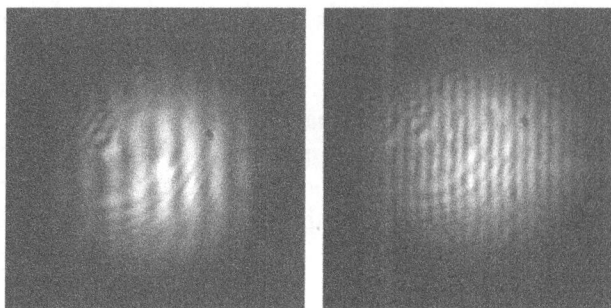

Figure 6.11. Near-field diffraction patterns at 0.83 and 1.82 MHz resonances.

6.2.2 Experimental suggestions

The velocity of sound in water is of order 1500 m s^{-1}. Suitable acoustic frequencies are in the region of 1–3 MHz, for which the wavelength in water is of order 1 mm. To get good diffraction patterns, the field of illumination should contain at least 10 periods of the standing wave. Since Λ can be measured using the diffraction pattern and the frequency is known, the velocity of sound in the water can be measured. The water can be replaced by other liquids, and also the temperature dependence can be investigated. Regarding the acoustic resonance in the cell, the refractive index amplitude of the standing waves can be estimated by measuring the values of the series of Bessel functions. This can be investgated as a function of the deviation of the acoustic frequency from the resonant value by comparing the intensities of the diffraction orders in figure 6.10(a) and (b), for example.

6.3 Berry's geometric phase

Berry's phase, γ, [6, 7] is a phase additional to the dynamic phase of a propagating wave which arises when the wave propagates along a route which is topologically equivalent to a helix. It is related geometrically to the propagation route. This can be seen most clearly when the propagation direction is drawn as a unit vector, which then traces a locus on the unit sphere as one progresses along the route. If the initial and final directions are the same, the Berry phase γ is equal to the solid angle Ω that the locus subtends at the centre of the sphere.

6.3.1 Berry's phase in an optical fibre

In the case of propagation along a single-mode fibre, the route is clearly defined geometrically by the route of the fibre, and so the value of γ can be determined from this. If the wave entering the fibre has a given polarization direction, it was shown [8] that the polarization vector at the exit from the fibre has rotated by angle γ, and this angle can be measured uniquely when the entrance and exit sections of the fibre are parallel (figure 6.12). We say 'uniquely' because if the two were not parallel, then the measured polarization angles would not be in parallel planes, nor would the solid angle on the unit sphere be defined uniquely since the locus would not be closed.

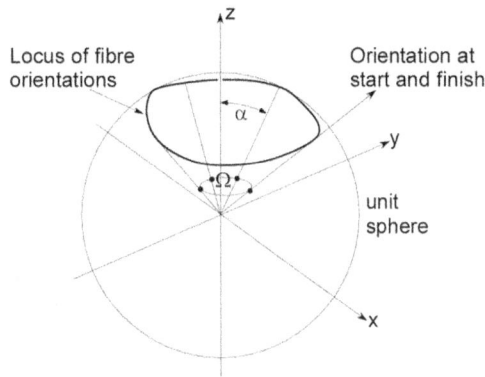

Figure 6.12. Locus of fibre orientations plotted on the unit sphere.

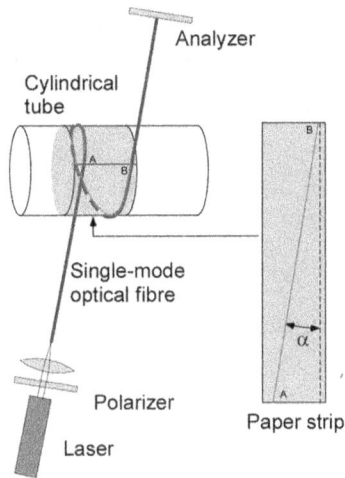

Figure 6.13. Fibre wrapped around the tube.

In the simplest experiment of this sort, based on Tomita and Chaio [8], we use a coiled single-mode optical fibre to create a Berry phase (figure 6.13). We prepare a paper strip with a diagonal line drawn on it and wrap it round a plastic tube of about 8 cm diameter. The diagonal line traces a helix at angle α to the axis of the tube. The fibre is then wrapped around the tube, following the line, so that its entrance and exit sections are parallel. A single loop is sufficient; the solid angle subtended by the route is then $\Omega = 2\pi(1 - \cos \alpha)$, and γ is equal to this. Clearly, if the fibre lies in a plane, whatever its route may be, $\alpha = 0$ and $\gamma = 0$. However, for other values of α, non-degenerate values of γ are obtained. Although the theory shows that γ is independent of the wavelength, it is convenient to use a laser for this experiment because of the need to focus the light onto the end of the fibre (which is probably the most difficult part of the experiment!). An assembly which is essentially the same as the laser's spatial filter (section 1.2) can be used for this, employing a microscope objective with magnification about 20 so as not to exceed the NA of the fibre. In the

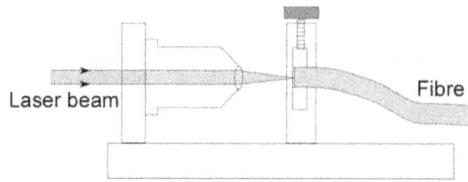

Figure 6.14. Focusing the laser beam onto the end of the fibre.

assembly, the pinhole is replaced by a disc of the same diameter as the pinhole mount, with the end of the fibre cemented into its centre. The x–y–z controls of the spatial filter are then used to bring the fibre end to the focus of the laser (figure 6.14). A polarizer precedes the focusing unit.

The experiment then consists of measuring the rotation of the angle of polarization as a function of α, which defines the solid angle $\Omega = 2\pi(1 - \cos \alpha)$. In the paper of Tomita and Chaio [8] in which this experiment was first reported, it was emphasized that the fibre must be of high quality so that it does not rotate the plane of polarization due to imperfections. We have found this not to be a problem with commercial communication fibres, but in any case, the fibre should be checked before starting the experiment by laying it on a planar surface and confirming that γ mod $(2\pi) = 0$. If this is not exactly so, the difference between the measured polarization angles in the planar and helical forms will presumably still give the value of γ.

6.4 Spatial coherence function: measurement and interpretation

The basic concept of coherence between two sources of light is an answer to the question: can light from the two sources interfere in a given observation plane to give stationary interference fringes, and what is the contrast obtained? In order that stationary interference be possible, the two sources must have a common origin. If that origin is an ideal laser, then light waves from the two sources will always be able to interfere. In the case where the origin is a conventional light source, interference can be observed only if the origin source is small enough. This effect was used, for example, as a method of measuring the size of stars, starting with work by A A Michelson in 1900 [9]. Since then, several large scale astronomical interferometers have been developed to create detailed stellar images, using spatial coherence measurements [10].

6.4.1 Measuring the spatial coherence function using Young's fringes

The contrast of the interference fringes measures the absolute value of a quantity called the 'coherence function', $\gamma(\tau, \vec{r})$, which is in general a function of relative time delay, τ, and vector separation, $\vec{r}$, between the two sources of the interference. In the basic experiment, we use a small conventional quasi-monochromatic source origin to illuminate an experimental plane, in which we want to find the spatial coherence function, $\gamma(0, \vec{r})$ when $\tau = 0$, i.e. when the optical paths from the origin to the point of interference via the two sources are equal. In practice, '$\tau = 0$' means that τ must

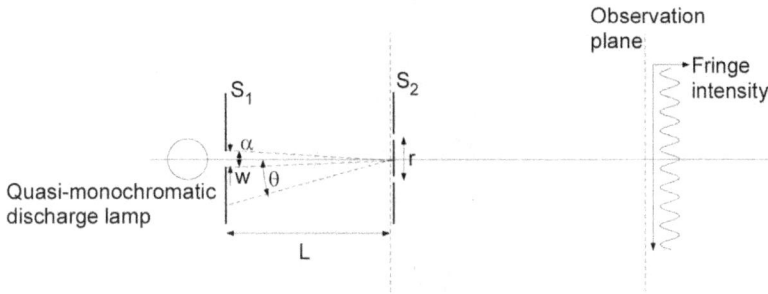

Figure 6.15. Basic spatial coherence experiment.

be smaller than the inverse of the spectral band-width of the source, for the measured value of γ to depend on r only. This experiment, which first investigated spatial coherence, was done by Thompson and Wolf [11] and is basically as shown in figure 6.15.

The theory of spatial coherence in 1D is based on the van Cittert–Zernike theorem [4, chapter X], which shows (paraxially) that $\gamma(r)$ in plane S_2 is the Fourier transform of $I(k_0\theta)$, which is the intensity in plane S_1. Clearly, in the example of figure 6.15, where α is the angular width of slit S_1 observed from the plane of S_2, $I(k_0\theta) = \text{rect}(2\theta/\alpha) = \text{rect}(2k_0\theta L/k_0 w)$ and so $\gamma(r) = \text{sinc}(rk_0 w/L)$ which has its first zeros at $r = \pm\lambda L/w$. The fringes resulting from the pair of slits separated by r in plane S_2 disappear when r has values which are multiples of this. In Michelson's original stellar interferometer, α of the star Betelgeuse was of order 5×10^{-2} arcsec (2.5×10^{-7} rad), which meant that the zeros of $\gamma(r)$ were at multiples of $\lambda/\alpha = 4\lambda \times 10^6 \approx 2$ m.

Since the coherence function depends on the product rw, this particular experiment might be carried out more easily by using a constant value of r and varying the entrance slit width w. The fringe contrast is measured using a camera and suitable analysis of the fringes. Note that if the slit S_1 remains coaxial, i.e. opens equally on both sides, one can visualize the phase of γ which results in the central fringe being dark rather than light in alternate periods of the sinc function. Practically, consider the case $L = 500$ mm, $\lambda = 0.6$ μm and $r = 2$ mm. Then the zeros of $\gamma(r)$ are at multiples of $w = 0.75$ mm. A convenient way of making a variable width slit which remains symmetrical to the optical axis is to use a tapered slit which can be translated vertically, and crossed by a fixed horizontal slit (figure 6.16).

6.4.2 Measuring the spatial coherence function using a shearing interferometer

Another way to carry out the experiment, with a fixed width for the slit S_1, is to use a shearing interferometer. For example, a Sagnac interferometer (section 5.3) can be built with a shearing plate (figure 6.17). When the plate is rotated, it provides two superimposed images of the object plane which is illuminated partially-coherently by the source. A plate thickness of 10–12 mm is convenient. When the camera is placed at the convenient A output of the interferometer, the fringe contrast is not 100%

Figure 6.16. Variable-width symmetrical slit.

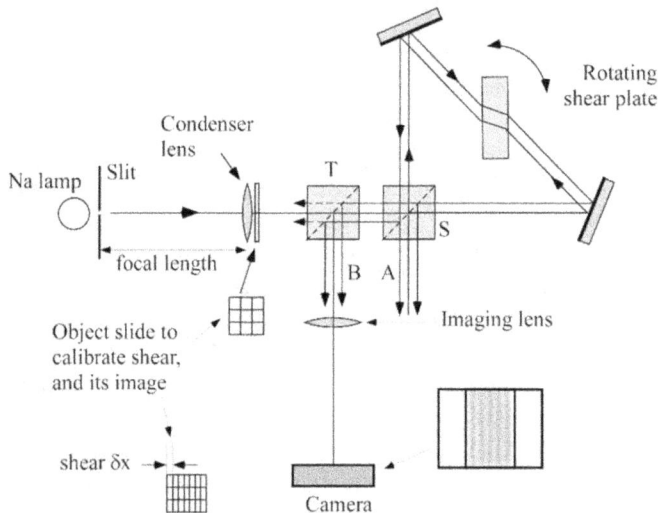

Figure 6.17. Shearing Sagnac interferometer.

even if the shear is zero, unless the beam-splitter S reflects and transmits exactly the same intensities, because the clockwise beam is transmitted twice and the anti-clockwise beam is reflected twice, and so their intensities are usually different. However, this problem is avoided if the less-convenient B output of the interferometer is used, because then both beams are reflected once and transmitted once (as in the Michelson interferometer). This requires addition of a second beam-splitter T which wastes three-quarters of the light, but this is not really a problem. The purpose of the condenser lens is to concentrate the light from the slit onto the imaging lens, so as to maximize the light throughput. The shear can be calibrated directly as a function of the angle of rotation of the shear plate by using a reticle which is placed temporarily in the object plane.

The results (figures 6.18 and 6.19) show clearly that the coherence function is the Fourier transform of the intensity profile of the slit source.

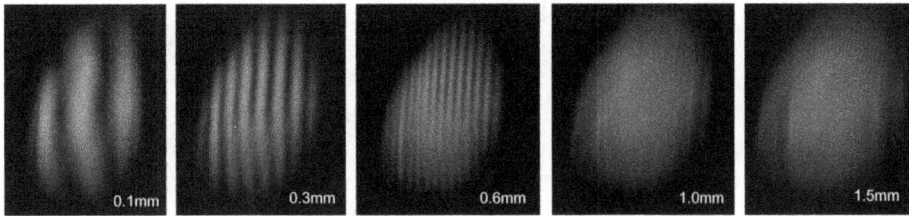

Figure 6.18. Typical fringes obtained with varying shear. The contrast is very poor at large shear!.

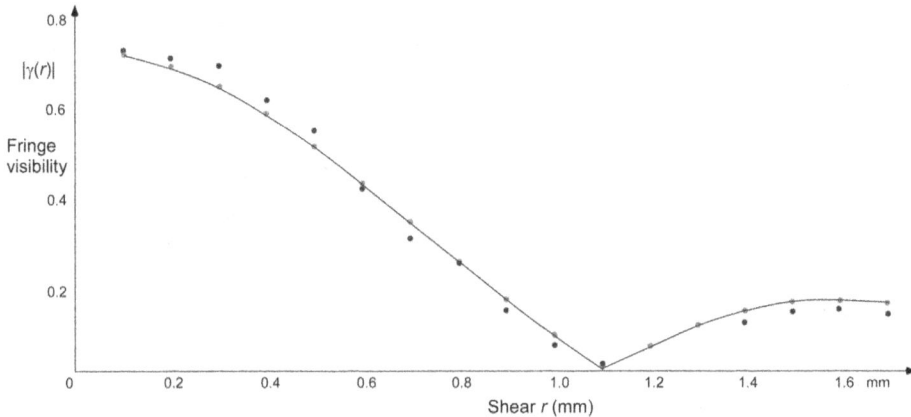

Figure 6.19. Measured contrast as a function of shear. The blue dots are measurements of the contrast and the red dots and their connecting line are the function $|\gamma(0)\mathrm{sinc}(\pi r/2.2 \text{ mm})|$, which is the Fourier transform of the illuminated slit.

6.5 Aperture synthesis

Aperture synthesis is the method which has been used for many years in radio astronomy, and more recently in optical astronomy, to achieve high resolution images of astronomical sources by combining coherently the signals from a number of relatively small apertures (telescopes) which are separated by large distances. For example, combining the signals from two very small apertures, with a variable time delay introduced, would give Young's fringes. The contrast of the fringes indicates the degree of spatial coherence between the light waves at the two apertures. Essentially, the apertures replace the two slits S_2 in figure 6.15. Now, as their relative positions change with time, due to rotation of the Earth or mechanical movement, the interference fringes between their signals then sample the spatial coherence of the source radiation sequentially at different vector separations. From these signals, an image of the source can be synthesized using the van Cittert–Zernike theorem [4, chapter X]. When this is done, we find that the angular resolution of the image constructed is inversely related to the largest separation between the detectors. The technique of aperture synthesis was first developed in radio astronomy by M Ryle, who received the Nobel prize for it in 1974. More recently, in 2019, signals from eight detectors at 1.3 mm wavelength on different continents have been combined to synthesize the image of a cosmic black hole (The EHT Collaboration [12]), whereas in optical stellar interferometry separations of hundreds of metres have been used [9].

Mathematically, the contrast of the interference pattern between the waves received by two telescopes whose positions are separated by vector $\vec{r}$ measures the coherence function $\gamma(\vec{r})$ of the wave field from the source on which the telescopes are focused. From the van Cittert–Zernike theorem, the incoherent source intensity as a function of position $\vec{\theta}$, i.e. the image $I(\vec{\theta})$, is the Fourier transform of $\gamma(k_0\vec{r})$, where k_0 is the wavenumber of the light detected by the interferometer. Measuring the absolute value $|\gamma(\vec{r})|$ is straightforward, even if it is technically challenging. For simple objects, such as double stars or stellar disks, the knowledge of $|\gamma(\vec{r})|$ is sufficient to reconstruct the image by an inverse Fourier transform. For more complicated situations, the phase of $\gamma(\vec{r})$ is required, and this is more difficult to measure.

In optical astronomy, the variation of vector $\vec{r}$ is achieved by two methods. One is to have two fixed telescopes and to use the rotation of the Earth to change the direction of the vector between them and its length projected on the plane normal to the direction of the observed star. From the contrast of the interference fringes obtained at wavelength $\lambda = 2\pi/k_0$ a sample of $|\gamma(\vec{r})|$ is acquired on a circular or elliptical locus of $\vec{r}$, limited by the hours of darkness. If three or more telescopes are used and interference fringes between all the pairs are recorded simultaneously, the relative phases of the values can be deduced (a technique called 'phase closure'). A second method uses moveable telescopes to change $\vec{r}$, so that $|\gamma(\vec{r})|$ can be sampled more completely in the plane of $\vec{r}$.

6.5.1 A laboratory aperture synthesis experiment

An experiment to illustrate the idea of aperture synthesis on an optical bench was described by Lawson *et al* [13]. There is an incoherent source consisting of a lamp behind a mask punctured by two or three small holes. Putting the mask in the focal plane of a lens creates an incoherent object at infinity and a filter defines the wavelength. The light is collected by two 'telescopes' represented by two small holes in a plate, and their interference pattern (Young's fringes) is observed using a CCD in the focal plane of a second lens. Now the plate is rotated in its own plane to represent the diurnal rotation of the telescopes and a series of interference patterns is recorded. Superimposing these patterns gives the synthetic image, because each one is a sine wave with amplitude $|\gamma(\vec{r})|$ and so the superposition does the inverse Fourier transform; this is not an accurate statement because the images show the intensity and not the amplitude, but it is sufficient for this demonstration. (A way to overcome this problem for the purpose of demonstration might to be to have an additional central hole with a larger diameter than D, so that the recording becomes a Gabor hologram, section 5.5.) The setup is shown in figure 6.20; the superposition can be seen by observing the CCD images on a monitor as the plate rotates, and using the eye to integrate them.

Some details are important. First, regarding the object mask, the diameter d of the individual holes in it must be sufficiently small that the illumination in the plane of

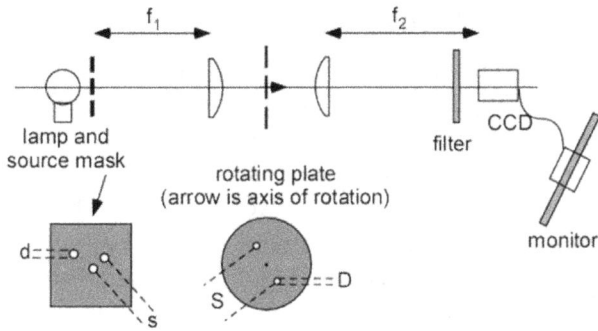

Figure 6.20. Setup for demonstrating aperture synthesis.

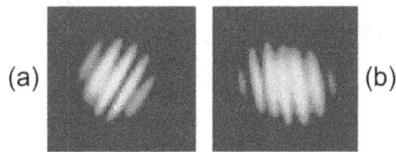

Figure 6.21. Typical interference patterns recorded by the setup of figure 6.20. Reproduced from [10] with permission of Cambridge University Press © 2006.

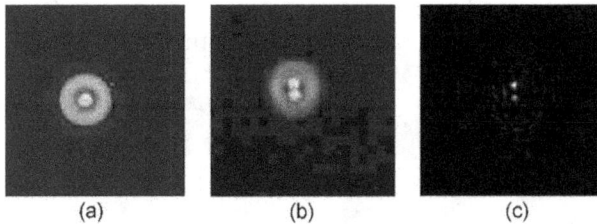

Figure 6.22. Synthetic images: (a) for one star and (b) a binary. (c) shows deconvolution of (b) using (a) as the point spread function. Reproduced from [10] with permission of Cambridge University Press © 2006.

the rotating plate has coherence distance greater than the distance S between the holes in that plate. This translates to $d < \lambda f_1/S$. In addition, the separation of the holes s must be large enough that the coherence function oscillates a few times in the distance S, i.e. $s > \lambda f_1/S$, otherwise the separate holes will not be resolved. Second, regarding the rotating plate, the diameter D of the holes in it must be small enough, compared with their separation S, to give rise to an interference pattern with several fringes, i.e. $S \approx 10D$. Finally, the focal length of the imaging lens must be such that the interference pattern fills much of the CCD, i.e. $f_2\lambda/D$ is about half of the CCD dimension. Figure 6.21 shows interference patterns from a binary source with the rotating plate in two orientations and figure 6.22 the synthetic images for a single star and a binary. One can see that the limited sampling of the Fourier plane leads to a distorted image, but in this simple case the images of the two stars can be recovered by Wiener deconvolution (figure 6.22(c)).

6.6 Gouy phase shift through a focus

When a wave with a spherical wavefront goes through a focus at its centre of curvature, its phase changes by π in addition to the phase it acquires from its propagation. This is called the 'Gouy effect' [14], and can be pictured very easily in the scalar wave approximation since a focused spherical wave converging on $r = 0$ has the form $E(r) = \frac{1}{r} \exp(ikr)$, so that as we go through the focus, the $1/r$ term changes sign abruptly. However, this singularity is unphysical. In reality, when a focused wave of finite extent, modeled as a Gaussian beam with amplitude proportional to $\exp[-r^2/2\sigma^2]$, goes through its focus, a focal spot of finite size results [15]. This is called a 'waist'. In this case, the same phase shift occurs, but it now takes place continuously within the waist region[1]. The region is defined by the Rayleigh length z_R, which is the distance from the waist to the plane where the beam has expanded to twice its minimum width (figure 6.23).

6.6.1 Experimental setup

The Gouy effect can conveniently be demonstrated using a Michelson interferometer in which one interferometer mirror, instead of being a plane mirror, is a concave spherical mirror (figure 6.23). An incoming plane wave is converted by this mirror into a spherical wave which converges to the waist. When the plane wave and the converging wave interfere, a set of circular fringes is obtained on a camera. The mirror positions can be adjusted to give a central dark spot (figure 6.23(b); zero phase difference) when the camera is placed a distance considerably greater than z_R

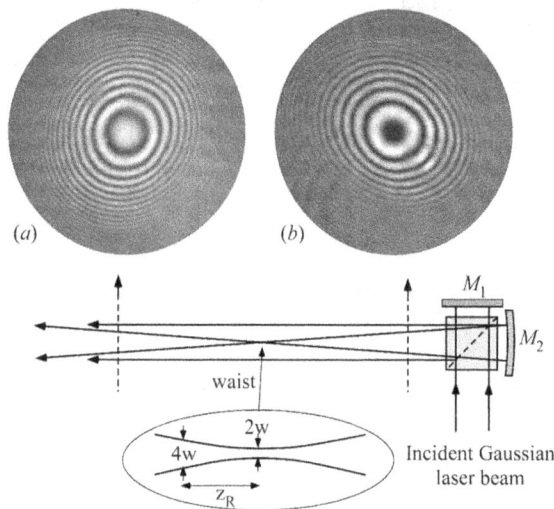

Figure 6.23. Experiment to show the Gouy phase shift. Reproduced from [16] with permission of Cambridge University Press © 2011.

---

[1] For a (n,m) laser mode, the Gouy phase change is $(1 + n + m)\pi$.

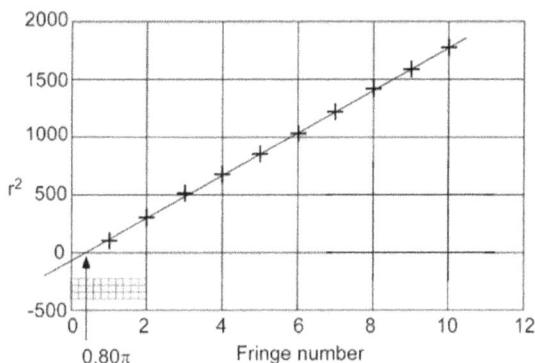

Figure 6.24. Plot of (fringe radius)^2 versus ring number for figure 6.23(a). The intercept on the x-axis gives the phase at the centre.

before the waist; this changes to a bright spot (figure 6.23(a)) at a distance greater than z_R after the waist, and indeed, the continuous phase change can be tracked in this way by translating the camera continuously through the focal region. The phase at the centre of the interference pattern may not be exactly 0 or π, but a simple plot of r^2 as a function of the ring number allows the phase at the centre to be determined accurately (figure 6.24). However, in order to get good results, the Michelson interferometer has to be very stable, so it is recommended to construct the interferometer on its own small table so that mechanical stresses resulting from movement of the camera affect the exact relative positions of the mirrors negligibly.

Experimental conditions might use a Gaussian laser beam with $\sigma = 5$ mm, focused by a mirror with focal length $f = 500$ mm. This gives a waist (Fraunhofer diffraction pattern) with Gaussian parameter $w = \frac{f}{k_0\sigma} = \frac{\lambda f}{2\pi\sigma} \approx 10$ μm. The waist region has length $z_R \approx wf/\sigma = 1$ mm, and the Gouy phase shift will be complete within a distance of a few z_R on each side of the focus. In order to get interference fringes with good contrast, one can move the camera to sufficiently large distances from the waist that the amplitudes of the two waves are approximately equal.

6.6.2 Two questions for investigation

What is the Gouy phase shift when a cylindrical wave goes through a *line* focus? If the beam profile is not Gaussian, does this modify the phase shift at the focus? Can you confirm your answers experimentally?

6.7 Optical vortices

Suppose that we have a transparent plate whose thickness $d(x, y)$ is a function of position. A plane wave is incident on the plate. The transmitted wave will have a position-dependent phase $\phi = k_0(n - 1)d(x, y)\cos\beta(x, y)$ relative to the incident wave, where n is the refractive index of the plate and β is the angle of refraction of the light in the plate. When we observe the interference pattern obtained in an interferometer when this transmitted wave interferes with a reference wave derived

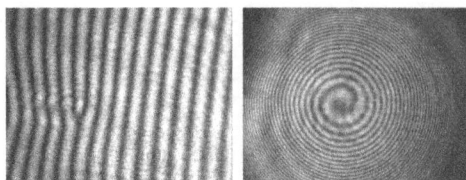

Figure 6.25. Interferograms of a vortex wave with $m = 1$ interfering with a plane wave and a spherical wave, respectively.

from the incident wave, we expect to see a set of continuous interference fringes representing the contours of $\phi(x, y) = 2\pi m$ where m is a positive or negative integer. If $d(x, y)$ and $\beta(x, y)$ are continuous functions, adjacent fringes must be separated by $m = \pm 1$. So what do we make of a situation in which the interference pattern has the form of a spiral, or contains a dislocation? In the spiral case, all the fringes apparently have the same value of m, and in the dislocation case two fringes which are adjacent on one side of the dislocation centre sprout intermediate fringes on the other side. Laboratory photographs of two such cases are shown in figure 6.25.

The waves shown in the figure are produced by a medium in which d is constant but $\beta(x, y)$ is not continuous[2], so that there is nothing wrong with the statement above. Waves having phase distributions with these spiral characteristics have an interesting property which is worth understanding and has some important applications: they carry angular momentum. They are called 'optical vortices'. The trick is to ensure that ϕ has a discontinuity of $2m\pi$ at all radii along the radial line $x = 0$, $y > 0$, which is achieved by a step in β along this line.

An optical vortex [17] is described by a wave propagating in vacuum along the z-axis with phase $\phi(r, \theta, z) = m\theta + k_0 z$ in a given plane z. Notice that we are now using cylindrical coordinates around the z-axis. Clearly there is a singularity along the axis of propagation $r = 0$; since β and hence ϕ can not be defined on this axis, the amplitude of the wave must be zero there, at all z. If we look at a localized region of the wavefront (ϕ = constant), at non-zero r, the wavefront is not normal to the axis, but has a gradient in the tangential direction, $\frac{\partial \phi}{r\partial \theta} = \frac{m}{r}$ so that locally the wave is traveling tangentially at angle ψ where $\tan \psi(r) = \frac{m\lambda}{2\pi r}$. In the quantum description, where the wave is replaced by an ensemble of photons each having momentum $\frac{h}{\lambda}$, the tangential component of the momentum is $\frac{h}{\lambda} \tan \psi = \frac{h}{\lambda} \frac{m\lambda}{2\pi r} = \frac{m\hbar}{r}$. This gives the photon angular momentum $r \times \frac{m\hbar}{r} = m\hbar$. Now a photon also has spin resulting from circular polarization, which can be either $\sigma = \pm 1$, depending on the sense of the polarization. This adds angular momentum $\hbar \sigma$, so that in total the angular momentum per photon is $(\sigma + m)\hbar$ in an optical vortex. Since the vortex angular momentum is added to the spin, it is often called 'orbital angular momentum' by

---

[2] Actually, it is $\cos \beta$ which should not be continuous. If the plate as described is placed symmetrically in the beam, so that angles vary from $-\alpha/2$ to $+\alpha/2$ the plate does not make a spiral beam. It must be mounted asymmetrically so the angles 0 to α are sampled.

analogy with that of an electron in an atom. However, it is questionable whether a single photon in a vortex has such an angular momentum, since there are no experiments in which a single such photon has contributed such angular momentum when exciting an atom, thus possibly allowing a forbidden atomic transition; but this is still an open question. The description is useful when considering an ensemble of many photons; then, if the wave is linearly polarized (a statistical ensemble of $\sigma = \pm 1$), the vortex contributes to a beam of power W a flow of angular momentum $\frac{W}{\hbar\omega} \times m\hbar = \frac{mW}{\omega}$. This result does not depend on quantization and can also be derived from Maxwell's equations, but is easier to follow when the photon concept is used.

6.7.1 Interference patterns

When two plane waves interfere with an angle α between their directions of propagation, straight-line fringes are created with distance $\lambda/[2 \sin \alpha/2] \approx \lambda/\alpha$ between them. If one of the waves has a vortex, the phase ϕ replaces that of that wave and the result is

$$E = E_0 \exp(i\alpha x) + E_0 \exp(im\theta);$$
$$I = |E|^2 = 2E_0^2[1 + \cos(\alpha x - m\theta)],$$

where $\theta = \arctan(y/x)$. This shows a dislocation at $r = 0$ (figure 6.25(a)). When a plane wave interferes with a diverging or converging wave, given by $E = E_0 \exp(i\gamma r^2)$, circular interference fringes are observed. If the plane wave is replaced by a vortex wave travelling along the same axis as the vortex wave, spiral interference fringes are observed. This occurs because as we go round a complete axial circle at constant radius, the phase of the vortex wave increases by $2m\pi$ and so a path difference of $m\lambda$ is added (figure 6.25(b)):

$$E = E_0 \exp(i\gamma r^2) + E_0 \exp(im\theta);$$
$$I = |E|^2 = 2E_0^2\left[1 + \cos(\gamma r^2 - m\theta)\right].$$

6.7.2 Creating vortex waves

There are several ways of producing vortex waves in the laboratory. The most commonly used one creates the phase pattern $\phi = m\theta$ by means of a liquid crystal spatial light modulator (SLM) in which the optical thickness of each pixel is determined by a computer-controlled voltage. These units are quite expensive, but very versatile. A very simple method [18], which was used to produce the examples shown here, is a thin Perspex plate with a radial crack in it (figure 6.26). The plate was made by cutting a sheet of Perspex about 3 mm thick with a guillotine, which is sufficiently crude as to create many cracks normal to the cut edge. A suitable disc about 30 mm diameter was located with a crack from its edge to its centre, and this was cut out and mounted in a frame with three pairs of opposed screws, as shown in figure 6.26(b), which allowed the disc to be distorted at will. The distortion was

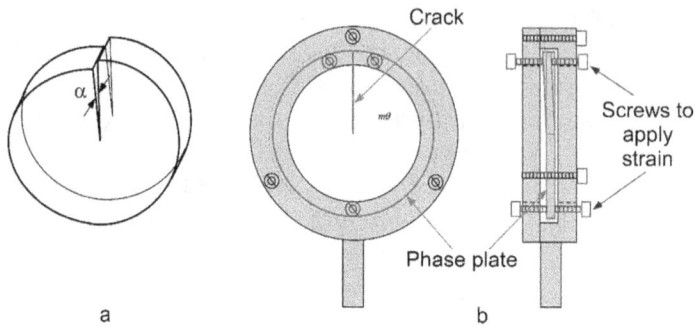

Figure 6.26. Mechanical construction of the phase plate.

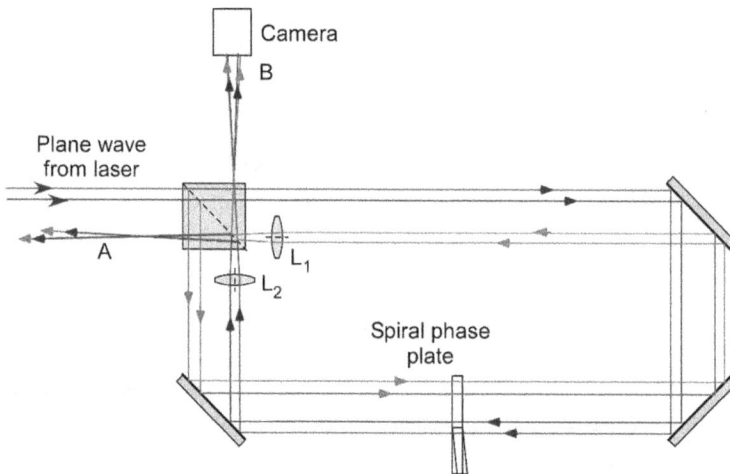

Figure 6.27. Rectangular Sagnac interferometer used to image the phase distribution of the exiting light wave. The clockwise beam is coloured blue, and is centred on the crack termination, while the anti-clockwise beam goes through a 'neutral' part of the plate and is coloured red. The camera is shown in exit B, but better contrast may be obtained at A. The lens positions in x, y and z can be controlled.

applied while the disc was mounted in the interferometer shown in figure 6.27, until a dislocation pattern was seen and the crack essentially vanished from sight.

The interferometer used was a rectangular Sagnac (see section 5.3) which has the advantage of being very stable even when forces are applied to the elements during observation. The construction shown in figure 6.27 allows observation in either the A or B exits: the A (symmetrical) exit shows better contrast, but the B exit is more convenient. The two lenses are used to control the interference patterns. L_2 produces an image of the wavefront exiting from the plate (it is best focused a short distance away from the plate so as to reduce sensitivity to defects on the Perspex surface), while L_1 produces a reference wave which has also passed through the plate, so the phase differences are close to zero. If both lenses are at the same distance from the camera (as seen through the beam-splitter), the waves reaching the camera have the same curvature and so the interference pattern shows directly the phase spiral

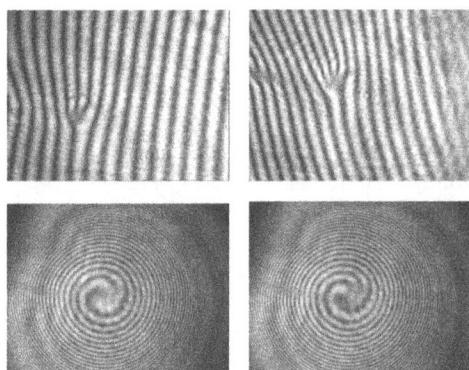

Figure 6.28. Interferograms for $m = 2$ (left) and $m = 3$ (right).

introduced by the plate. This is easier to appreciate when there is a small angle between the two wavefronts, so that in the absence of the plate distortion straight line fringes would be seen. This is attained by lateral movement of L_1, which does not move the wavefront image, but produces a controllable angle between the fields. The resulting dislocation image is shown in figure 6.25 (left). On the other hand, if both lenses are coaxial to their beams, but L_1 is moved axially, a curvature difference between the object and reference waves results which provides a circular fringe background and the spiral pattern shown in figure 6.25 (right). In figure 6.28 we show interferograms for the cases $m = 2$ and $m = 3$.

When a spiral phase has been constructed and tested, the following experiments with it can be carried out. First, the good region of the plate should be isolated by means of an iris diaphragm centred on the singularity. Then, its Fraunhofer diffraction pattern can be investigated; this should have the form of $J_{m+1}(kr)/kr$, which has a zero at its origin; this property is used in the STED technique for super-resolution in fluorescent samples (Nobel prize, 2014). In another experiment, the spiral phase beam is focused into water containing microscopic particles which are observed under a microscope, and the particle can be seen to rotate due to absorption of angular momentum from the beam. This is called an 'optical wrench'!

References

[1] Bose J C 1897 On the influence of thickness of air-space on total reflection of electric radiation *Proc. Roy. Soc.* A **62** 300
[2] Zhu S, Yu A W, Hawley D and Roy R 1986 Frustrated total internal reflection: a demonstration and review *Am. J. Phys.* **54** 601
[3] Vörös Z and Johnsen R 2008 A simple demonstration of frustrated total internal reflection *Am. J. Phys.* **76** 747
[4] Born M and Wolf E 1999 *Principles of Optics* (Oxford: Pergamon) ch XII, X
[5] Ribak E 2001 Harnessing caustics for wave-front sensing *Opt. Lett.* **26** 1834
[6] Berry M V 1984 Quantal phase factors accompanying adiabatic changes *Proc. R. Soc. Lond.* A **392** 45
[7] Berry M V 1987 Interpreting the anholonomy of light *Nature* **326** 277

[8] Tomita A and Chaio R Y 1986 Observation of Berry's topological phase using a coiled optical fibre *Phys. Rev. Lett.* **57** 937

[9] Michelson A A 1927 *Studies in Optics* (New York: University of Chicago Press) reprinted (1995) by Dover

[10] Labeyrie A, Lipson S G and Nisenson P 2006 *An Introduction to Optical Stellar Interferometry* (Cambridge: Cambridge University Press)

[11] Thompson B J and Wolf E 1957 Two-beam interference with partially coherent light *J. Opt. Soc. Am.* **47** 895

[12] The Event Horizon Telescope Collaboration *et al* 2019 First M87 EHT results *Astrophys. J. Lett.* **875** L1–6

[13] Lawson P R, Wilson D M A and Baldwin J E 2003 Desktop interferometer for optical synthesis imaging *Proc. SPIE* **4838** 404

[14] Gouy L G 1890 Sur une proprieté nouvelle des ondes lumineuses *C. R. Acad. Sci., Paris* **110** 1251

[15] Saleh B E A and Teich M C 1991 *Fundamentals of Photonics* (New York: Wiley)

[16] Lipson A, Lipson S G and Lipson H 2011 *Optical Physics* 4th edn (Cambridge: Cambridge University Press)

[17] Padgett M and Allen L 2000 Light with a twist in its tail *Contemp. Phys.* **41** 275

[18] Rotschild C, Zommer S, Moed S, Hershcovitz O and Lipson S G 2004 An adjustable spiral wave-plate *Appl. Opt.* **43** 2397

IOP Publishing

Optics Experiments and Demonstrations for Student Laboratories

Stephen G Lipson

Chapter 7

Optics of materials

7.1 Interferometric measurement of the refractive index of a gas

This experiment can be done conveniently using a Jamin interferometer, although other interferometers might also be employed. The Jamin interferometer (figure 7.1) has the advantage of producing stable fringes, like the Sagnac, even if it is not mounted on a floating table, and measuring the refractive index of a gas such as air, CO_2 or He is a good example of its application. The interferometer measures the change in optical path through the gas as its pressure P is changed from vacuum to atmospheric pressure. The dimensions of the setup needed are quite convenient. The refractive index of air at atmospheric pressure and room temperature is 1.000 29, so that in a tube of length L the change in optical path between vacuum and atmospheric pressure is $L(n - 1) = 2.9 \times 10^{-4} L$. For $L = 100$ mm, this is 29 μm, which is about 50 wavelengths. In an experiment, the fringe shift can therefore be measured as a function of pressure to determine $(n - 1)$ to an accuracy of about 1%. Since both interfering waves travel the same distance, apart from the difference in optical path between the gas in the tube and in the parallel air space, the laser required for the experiment could be a laser diode, provided that its line-width is considerably less than 1/50 of its wavelength. Using two or three lasers with different wavelengths, the Abbe dispersion index (section 2.1) can also be estimated. Since the glass plates needed for the interferometer act as beam-splitters, they must have thickness approximately equal to twice the separation between the two interfering beams (prove this); they do not need to be coated with a reflective layer since both beams suffer two reflections before they interfere, so the fringe contrast is high, even if much light is lost. Also, since the laser beam is not expanded, the optical flatness of the plate surfaces does not need to be exceptional, and commercial window glass can be used. The diverging lens in the figure expands the fringe pattern, so that counting the fringes and even interpolating their positions to about 1/4 fringe, becomes easier.

doi:10.1088/978-0-7503-2300-0ch7

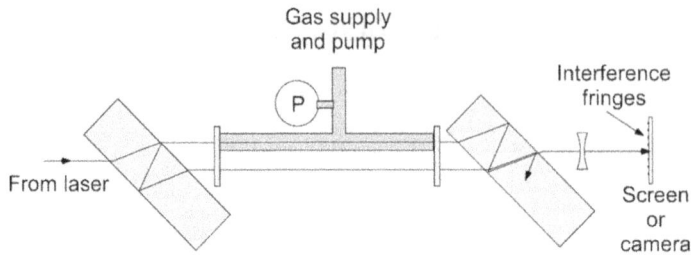

Figure 7.1. Jamin interferometer used to measure the refractive index of a gas.

7.2 Anisotropic materials: interference figures of uniaxial and biaxial crystals

7.2.1 Basic description of birefringent crystals in terms of the refractive index surface

A convenient way of describing the properties of an anisotropic crystal transmitting light is in terms of its refractive index surface[1]. We assume here that an optical material is magnetically isotropic, which means that the magnetic field vectors $\vec{H}$ and $\vec{B}$ are parallel in all cases; this may not be true for some exotic meta-materials. This surface is derived from the dielectric constant tensor, expressed geometrically as an ellipsoid (also called 'optical indicatrix'). In general, because an ellipsoid has three different principal values, a wave in a certain direction $\hat{k}$ has different refractive indices for two orthogonal principal polarizations, and it propagates as the superposition of two characteristic plane waves with these polarizations. The two indices can be plotted as radial functions of the propagation direction $\hat{k}$, generating a double-valued surface. In general, the inner and outer branches of this surface touch at four points (two pairs of diametrically opposed points) which define two 'optic axes', and the crystal is called 'biaxial'. The two characteristic waves propagating along the optic axis have the same velocity, because their refractive indices are equal. If the crystal has a degree of symmetry, the two optic axes may be degenerate and there is a single optic axis, giving a 'uniaxial' crystal. A photograph of a model of the refractive index surface for a biaxial crystal is shown in figure 7.2. By definition, the $\vec{k}$ vector of a wave in the crystal is the radius vector of the surface; its length is $\hat{k}_0 \cdot n(\hat{k})$ and the Poynting vector (ray direction) $\hat{S}(\hat{k})$ is the normal to the surface. If $\hat{k}$ and $\hat{S}$ are parallel the ray is called an 'ordinary ray' and if not, it is an 'extraordinary ray'. It is easy to visualize the polarization of the extraordinary wave, because $\vec{H}$ and $\vec{B}$ are normal to both $\hat{k}$ and $\hat{S}$; the polarization of the ordinary wave is then normal to that of the extraordinary one. When a wave travels along the optic axis of a birefringent crystal one can see from the figure that although $\hat{k}$ is well defined, $\hat{S}$ is not, since it is normal to the refractive index surface. This gives rise to

---

[1] Not all optics textbooks use the refractive index surface as a basis for understanding waves in anisotropic crystals. I believe that this is the easiest approach to the subject, and is an analogy to the use of the Fermi surface in solid-state physics to understand electron dynamics. The method is explained in detail in reference [1], chapter 6.

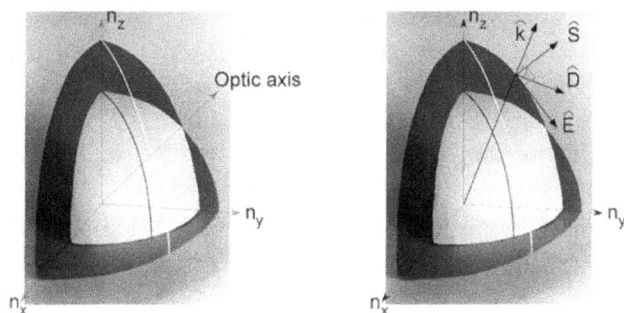

Figure 7.2. One octant of a model showing the double-valued refractive index surface of a biaxial crystal. The optic axis is the direction where the inner (white) and outer (black) branches touch. The right figure shows construction of the **D** and **E** vectors for a given **k** when there is no magnetic anisotropy: **S** is normal to the outer surface where **k** intersects it, and **B** is normal to **k** and **S**. Then **D** is normal to **k** and **B**, and **E** is normal to **S** and **B**.

Figure 7.3. A $NaNO_2$ crystal, as grown, used as a prism to refract an unpolarized beam. The beam is incident from the top left, creating a reflected wave to the right, and two refracted waves having orthogonal polarizations towards the bottom.

the phenomenon of *conical propagation* of the Poynting vector, which has a singularity at that point.

7.2.2 Uniaxial and biaxial crystals

Several birefringent crystals can easily be grown from solution, using standard methods. These include KDP (potassium di-hydrogen phosphate, KH_2PO_4) and sodium nitrite ($NaNO_2$). Alternatively, crystals such as calcite (uniaxial) and mica (biaxial) are easily available. Notice that the crystals have low symmetry, and their unit cells do not have right-angled corners like NaCl; this lack of high symmetry is necessary for them to be birefringent. A piece of extruded plastic sheet, such as an overhead projector slide, behaves as a birefringent crystal since one direction (the thickness) is compressed, one direction (the length) is stretched and the third dimension is unstressed. Some straightforward experiments using such crystals are as follows:

    a. *Two refractive indices.* Using the corner of a crystal as a prism to deflect a laser beam, the two different refractive indices n_o and n_e are evident as two emerging beams, having orthogonal polarizations (figure 7.3).

b. *Interference figure of a thin birefringent plate.* Using a parallel-sided crystal plate sandwiched between crossed polarizers and illuminated by a convergent beam of light, its interference figure can be seen in the far field. The interference figure is the transmitted intensity as a function of angle of incidence (in two dimensions), which results from interference between the parallel propagation of the two waves with refractive indices n_{inner} and n_{outer}, when the incident wave is polarized at an angle between those of the two characteristic waves. Simple ways of doing the experiment are to focus a laser beam onto the sandwich and to observe the light intensity on a distant screen or, if the crystal is big enough, to put the sandwich in front of your eye and look at a cloudy sky. The figures are characteristic of uniaxial and biaxial crystals (figure 7.4). Figure 7.4(c) used a piece of overhead projector slide [2], and the angular range of observation was sufficiently large to show both optic axes; one axis is detailed in figure 7.4(d). It is an interesting exercise to determine the differences between the three principle refractive indices of the crystal from such a photograph (given the thickness and the angle calibration). If the illumination is broad band, the fringes are coloured, like in white-light interferometry, since the principal refractive indices are equal on the optic axis (figure 7.4(b)). The origin of the black cross in figure 7.4(a) and (b) and the black fringes through the optic axes in figure 7.4(c) and (d) can be understood by considering the polarizations of the ordinary and extraordinary rays in each direction of $\vec{k}$, which are parallel or perpendicular to the incident polarization in these directions.

c. *White light interference in a birefringent plate.* When a thin parallel-sided crystal plate (thickness d) is observed in an approximately collimated beam of white light between crossed polarizers, and the polarizer/analyzer axes are inclined with respect to the axes of the crystal, coloured images are formed

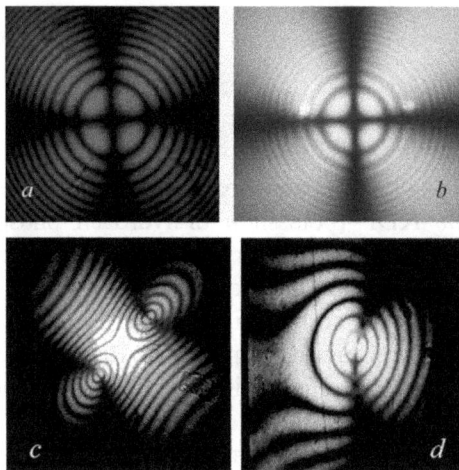

Figure 7.4. Interference figures: (a) and (b) uniaxial lithium niobate, in monochromatic and white light, respectively; (c) biaxial overhead slide in monochromatic light; (d) biaxial, centred on one axis. Reproduced from [1] with permission of Cambridge University Press © 2011.

by interference between the two waves travelling through the crystal. The wavelengths for which $(n_o - n_e)d/\lambda$ is an integer are blocked by crossed polarizer and analyzer, and when this quantity is an integer +1/2, they are transmitted with least attenuation. This can easily be observed using mica, or stretched transparent tape (figure 7.5). An important device which uses this effect is the *quarter-wave plate*, often constructed from a mica sheet, which has thickness such that $\frac{(n_o - n_e)d}{\lambda} = \frac{1}{4}$, so that the two principal waves E_x and E_y transmitted by it have phase difference $\frac{\pi}{2}$ and the resultant wave is circularly polarized. Note that mica is actually biaxial (see section 3.1.3), but it cleaves in such a way that the x and y axes are in the plane of the cleaved crystal, and z is normal to it. This is convenient because for a wave travelling parallel to z the two relevant refractive indices are n_x and n_y, which are very close to one another, so that d is quite large. Circularly polarized light is produced when the incident polarization is at 45° to the x and y axes (one sense of rotation) or −45° (the other sense).

7.3 Chiral materials: optical activity

In a medium which has chiral structure, the characteristic waves are circularly polarized. Such materials are called 'optically active'. The chiral structure may be oriented, as in quartz, which is basically uniaxial. However, when a wave propagates along the optic axis, the two characteristic circularly-polarized waves have slightly different velocities, so that the inner and outer parts of the refractive-index surface do not quite touch, as they do in an inactive uniaxial medium. Now, a linearly-polarized wave can be represented as the superposition of these two characteristic waves. As a result of the small difference between the refractive indices of these two waves, the direction of the polarization rotates as the wave propagates.

In solutions of chiral molecules, such as sugar solutions, the molecules have random orientations, so that the above effect occurs for propagation in any direction. The refractive index surface for such a medium has the form of two concentric spheres, with refractive indices n_l and n_r for left- and right-handed

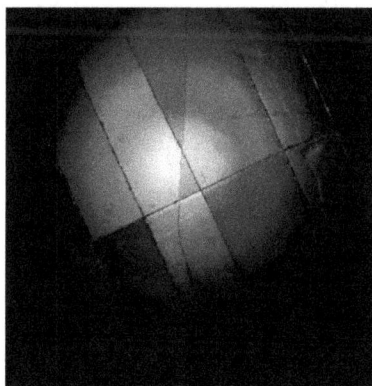

Figure 7.5. Some overlapping strips of sellotape observed between crossed polarizers.

circularly polarized waves, respectively. Then the polarization vector of a linearly-polarized propagating wave rotates continuously as the wave progresses. The distance in which the electric field makes a complete rotation is $z = \lambda/(n_l - n_r)$, so that the field is parallel or antiparallel to a given direction at intervals of $\Delta z = \lambda/2(n_l - n_r)$. If a container of sugar solution is viewed between two linear polarizers which are parallel, the transmitted light selects the colours satisfying this equation. If one of the polarizers is rotated, the colour changes continuously. The colours are not very pure, because the selection is equivalent to a very broad band filter. Corn syrup is a good medium for such experiments, having a very high fructose concentration, and Δz is of the order of a few cm.

An experiment which combines this effect with Rayleigh scattering uses corn syrup with a small concentration of scattering particles. Then, the scattering changes periodically along the axis of a polarized wave in the medium, as the polarization vector rotates. Using a HeNe laser, the period observed is about 8 cm.

7.4 Non-linear optics: second harmonic generation

Note that this experiment involves the use of a high power laser, and suitable precautions must be taken to avoid eye exposure.

When an electromagnetic wave travels through a transparent medium, its propagation can be described by considering how each molecule of the medium becomes polarized by the electric field of the incident wave, and how the resulting oscillating dipole moment of the molecule radiates. The sum of the transmitted and radiated waves describes the propagation of the light within the medium. This is one way of relating the polarizability of the molecules to the refractive index of the medium. It is like Huygens's construction, except that in this case the reradiation is a real phenomenon, and not just a hypothesis! If there is a phase delay due to a resonant mechanism in the response of the molecule, this also relates to an absorption index. The model assumes implicitly that the radiating waves are at the same frequency as the incident wave. However, if the intensity of the incident wave is very high, the dipoles may not oscillate in a sinusoidal manner, and the radiated waves may contain higher harmonics of the fundamental frequency. In that case, part of the power of the incident wave will be converted to a wave at a harmonic frequency, which is an integer multiple of the incident frequency. This experiment considers the generation of a second harmonic, at twice the incident frequency. Second harmonic generation is used commercially in green laser-pointers, which use essentially the same components as described below.

7.4.1 Phase matching

The most important issue in second harmonic generation is to ensure that the radiated waves interfere constructively, which is called *phase matching* [3, 4]. Since they are radiated in phase with the incident wave, and are travelling in the same direction, the waves will interfere constructively if the velocities of the two waves are equal, which is not generally the case because their wavelengths are different. However, dispersion in the crystal can be compensated by using the different refractive indices of waves with different polarizations. Let us consider a uniaxial crystal, whose optical properties are described by its refractive index surface, as described in section 7.2. This surface has two branches, which correspond to orthogonal polarizations for any given direction $\hat{k}$. In figure 7.6 we show a section of the refractive index surface for a negative uniaxial crystal, where the optic axis is in the z-direction and the two branches are the ordinary wave branch, which is a sphere (i.e. refractive index and wave velocity are independent of direction of propagation) and the extraordinary branch, which is a prolate spheroid (an ellipsoid like a Rugby football). The two branches touch along the optic axis. The sections are shown for frequency ω in red and for 2ω in blue, the difference in scale resulting from dispersion.

Type 1 phase matching of the incident and harmonic waves is achieved for propagation along the axis $\hat{k}_p$ which is the axis along which the refractive indices of the ordinary wave at frequency ω and the extraordinary wave at frequency 2ω are equal. This involves a substantial coupling mechanism between two orthogonal polarizations (the incident wave normal to the paper and the harmonic wave parallel to the paper in figure 7.6), which is a required property of the crystal being used.

Type 2 phase matching is appropriate when there is substantial coupling between waves with parallel polarizations. It is achieved by mixing the incident waves of both o- and e-polarizations using a polarizer at 45°, resulting in an amplitude-modulated wave polarized in that direction:

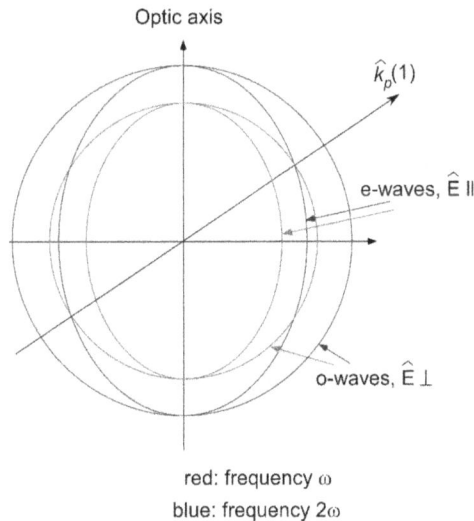

red: frequency ω
blue: frequency 2ω

Figure 7.6. Type 1 phase matching in a negative uniaxial crystal: the wave-vector direction $\hat{k}_p$, along which matching is achieved, is determined by the intersection of the extraordinary branch of the refractive-index surface at 2ω with the ordinary branch at ω.

$$E = E_0 e^{i\omega t}[\exp(ik_0 n_o s) + \exp(ik_0 n_e s)]/\sqrt{2}$$
$$= E_0 \sqrt{2}\, e^{i\omega t} \exp\left[ik_0 s(n_o + n_e)/2\right] \cos\left[k_0 s(n_o - n_e)/2\right]. \tag{7.1}$$

This wave has a carrier phase velocity $2c/(n_o + n_e)$ and effective refractive index $(n_o + n_e)/2$. By using the orientation of the crystal to adjust $n_e(\omega)$, this can be made equal to the harmonic extraordinary velocity $c/n_e(2\omega)$, as shown in figure 7.7.

The crystal being used will indicate which type of phase matching is appropriate, according to the values of the elements of the electric polarizibility tensor. If the tensor has strong off-diagonal components, the incident and radiated waves will be polarized orthogonally and type 1 phase-matching may be appropriate. If the tensor is essentially diagonal, then the incident and radiated waves have similar polarizations, so that type 1 phase-matching will not work. In that case type 2 has to be used since the matching is between two waves with polarizations at 45° to one another, and so the incident and radiated waves have non-zero parallel electric field components.

Crystals which can readily be purchased for this experiment are BBO (beta-barium borate, β-BaB$_2$O$_4$) transmission range 196 nm ~ 2200 nm, KTP (KTiOPO$_4$) transmission range 350 nm ~ 4500 nm. Moreover, KDP (KH$_2$PO$_4$) can quite easily be grown from solution in the lab and a suitable crystal plate made from it.

7.4.2 The experiment

The experiment uses a plate of potassium titanyl phosphate (KTP) crystal (type 2) with the optic axis in its plane and a Nd:YAG laser of wavelength 1064 nm polarized parallel or at 45° to the optic axis of the crystal, depending on the appropriate type of phase matching. If the laser output is polarized, a half-wave plate for its wavelength can be used to rotate the plane of polarization to the required direction

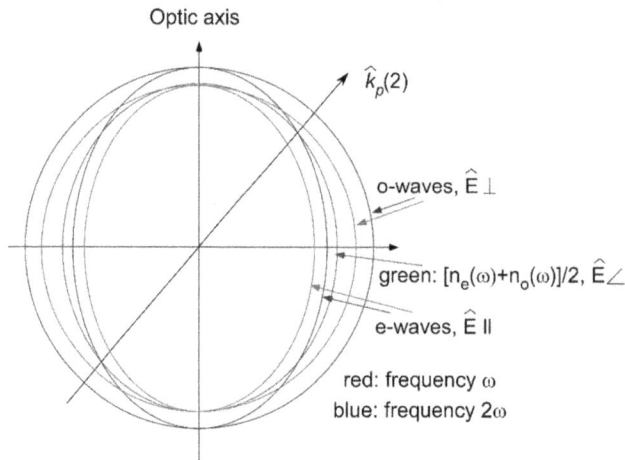

Figure 7.7. Type 2 phase matching: the wave vector is determined by the intersection of the average of ordinary and extraordinary branches of the refractive-index surface at ω (shown in green) with the extraordinary branch at 2ω.

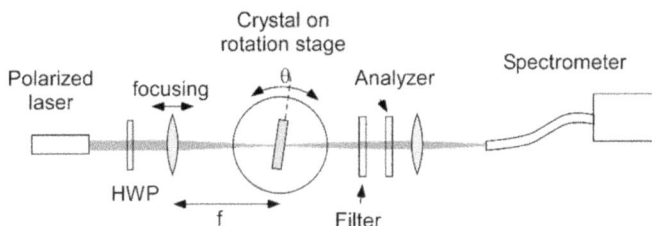

Figure 7.8. Layout of the optics. A polarized 1064 nm laser is focused into the crystal using a glass lens, through a half-wave plate which is used to vary the plane of polarization. The spectrometer is coupled to the beam via an optical fibre, and must span the wavelength range of at least 500–1100 nm. Both the focusing and the crystal rotation angle need micrometer control. The output of the spectrometer at 532 nm should be displayed prominently.

without losing intensity. The laser is focused into the crystal, so as to achieve maximum power density and the exiting light beam passes through a filter centred on the second harmonic wavelength 532 nm. The beam then goes through an analyzing polarizer and is focused onto an optical fibre which brings it to a spectrometer, so that the energy in the 532 nm wave can be measured (this avoids pollution by a small fraction of the incident wave which might pass the filter) (figure 7.8).

One experiment can measure the efficiency of the phase matching that can be achieved with the given crystal and optics. If there is a small error $\delta k = k(2\omega) - 2k(\omega)$ in the phase matching, the efficiency is reduced by partially destructive interference between the generated waves. Within the thickness t of the crystal the interference can be described by the function

$$E(2\omega) = \alpha E_0 \int_{-t/2}^{t/2} \exp(i\delta ks)ds = \alpha E_0 \, \mathrm{sinc}(\delta k t/2)$$

and the intensity is therefore proportional to $\mathrm{sinc}^2(\delta k t/2)$. Since the refractive index of an extraordinary wave is a function of the direction of propagation, δk can be varied by control of θ, the angle of incidence of the wave (figure 7.9). Another experiment can verify the dependence of the phase-matching on the direction of polarization ϕ of the incident wave, with respect to the optical axis of the crystal. If the polarization of the incident wave is rotated by 90°, in equation (7.1) the two interfering waves would be in antiphase instead of in phase, and the modulated sum would have sine instead of cosine, which would not change the outcome. So, as a result, the second harmonic output intensity varies as $\cos^2[2(\phi - \pi/4)]$. What angular variation would be expected for a type 1 crystal? Another experiment investigates the effect of focusing the incident wave. Using a shorter focal-length lens concentrates the power more in a smaller focal spot, but the length of the focal region (the Rayleigh length) is also shorter. In view of this, how does changing the focal length affect the efficiency?

If the spectrometer range goes down to 300 nm, it might be possible to detect a third harmonic. This harmonic is common to all crystals, but is probably very weak.

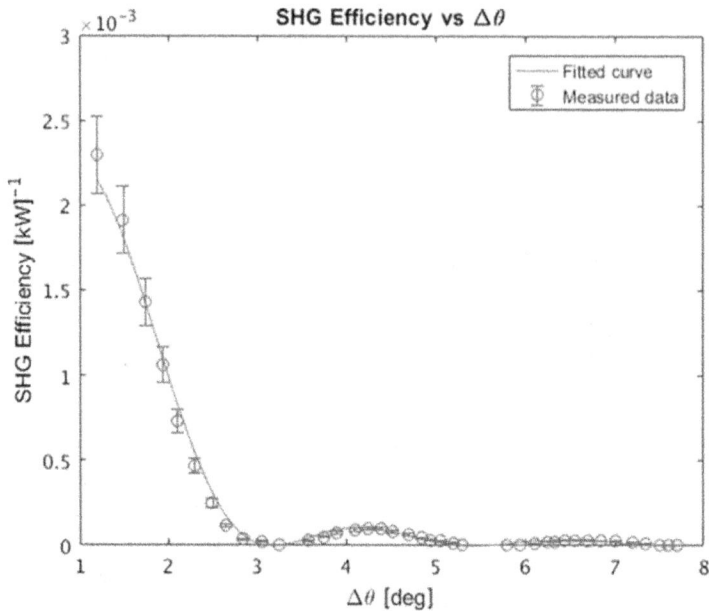

Figure 7.9. Second harmonic generation efficiency in KTP as a function of the angle of incidence, $\Delta\theta$ relative to the angle of phase-matching.

7.5 Surface plasmon resonance

Surface plasmon resonance is a very sensitive method for the detection of changes in the boundary conditions at a metal–dielectric interface. It is capable of detecting changes of the dielectric constant of the order of 10^{-6}, and has been developed as a biosensor, for example, in which interactions between bio-molecules adsorbed on a metal surface can be sensed in real time.

We know that in an optical fibre a light wave can be localized so that it propagates in a region of higher refractive index which is *surrounded* by a region of lower index. The question arises: is it possible to localize a wave on a *single* interface between two regions of different refractive indices? It turns out that this is possible, provided that one of the materials is a dielectric with a positive dielectric constant ε_d and the other has a dielectric constant ε_m which is complex and has a negative real part, a condition which is satisfied by metals. Such propagating waves are called 'surface plasmons' or 'surface polaritons'. Surface plasmons were discovered accidentally in the early 1900s by R W Wood when he was investigating metallized diffraction gratings inscribed on glass surfaces. The theory is fairly straightforward for a planar interface $x = 0$. We look for a solution of Maxwell's equations which is a wave propagating in the z-direction that has amplitude decaying exponentially in both $x > 0$ and $x < 0$, and satisfies the usual boundary conditions at the interface. The details are described briefly in reference [1], section 13.7, and more extensively by Maier [5]. It turns out that the solution must satisfy the following conditions regarding the wave polarization, the complex dielectric constants of the metal ε_m and the dielectric ε_d at the wave frequency:

1. The wave must have p-polarization (i.e. only E_y).
2. The metal must satisfy the condition: real $(\varepsilon_m) < -\varepsilon_d$.

The propagation vector k_{zp} of the propagating wave is then related to that of the incident wave in vacuum k_0 by the relationship $k_{zp} = k_0 \sqrt{\frac{\varepsilon_d \varepsilon_m}{\varepsilon_d + \varepsilon_m}}$. When condition (2) is satisfied, the real part of the square root is dominant, and so the wave has a long propagation distance.

Condition (2) limits the metals on which surface plasmons can propagate at an interface with air or glass ($\varepsilon_d = n^2 \approx 2.5$). For example Au ($\varepsilon_m = -8.9 + 1.2i$) is a popular choice, and Ag, Al, In and other metals also satisfy the requirements. When ε_m is known, the complex propagation vector k_{zp} gives the wavelength and decay length of the surface plasmon wave.

7.5.1 Observing the plasmons

To observe the plasmons, we must excite them by coupling them to a propagating probe wave which has the same frequency and propagation vector, in which case there is a resonance between them which transfers energy from probe to plasmon. Since from condition (2) real $(k_{zp}) > k_0$, this cannot be done by a free space wave, for which $k_{zp} = k_0 \cos \theta$, where θ is the angle of incidence of the wave. Three methods of coupling are commonly used: one uses a diffraction grating to add a constant $1/(\text{grating period})$ to the wave vector k_0, and the other two use the evanescent waves which propagate in the region outside the hypotenuse of a prism when the incidence is above the critical angle. Of these two, the more widely used is the 'Kretschmann configuration', in which a thin metal film is deposited on the hypotenuse of a glass prism and the surface plasmons are excited on the external metal surface (figure 7.10(a)) [6, 7]. This configuration has many applications, particularly as a biosensor, since the plasmon propagation is modified by attachment of biological molecules on the outer surface (changes in ε_d). The other one, the 'Otto configuration', uses the evanescent wave in the air-space adjacent to the prism hypotenuse to couple the probe wave to the metal (figure 7.10(b)), which can be a thin deposited film or a bulk sample, and will be a suggested experiment in this chapter. The Otto configuration has not been used widely because of the difficulty of controlling and measuring the separation d, but in the experiment described here the prism surface is

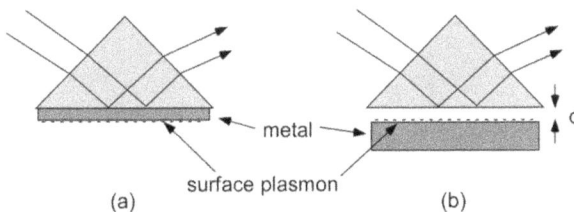

Figure 7.10. (a) Kretschmann configuration and (b) Otto configuration.

made slightly convex, so that the separation varies continuously with position, and the configuration of Newton's rings, applied to optical tunnelling in section 6.1, can be applied directly. The advantage of the Otto configuration is that it is not restricted to thin films of metal, but can be used with bulk specimens too.

7.5.2 Experiments using the Kretschmann configuration

In the Kretschmann configuration (figure 7.11) we have a plane wave incident from inside the prism at angle θ onto the hypotenuse, on which a thin film of metal has been deposited [6, 7], so that the wave number $k_z = k_0 n_d \sin \theta$. We observe the intensity of the p-polarized reflected light (figure 7.12) as a function of θ. As θ increases from below the critical angle, the intensity increases to a maximum at the critical angle. After that, at a certain angle the intensity drops sharply, reaching a minimum at the surface plasmon resonance, when k_z matches that of the surface plasmon, after which it rises again. If the film is too thin, the plasmons are not well defined and the resonance is poor; if the film is too thick, the light wave is attenuated on passing through the film and again the resonance is poor. The optimum film thickness of Au or Ag is about 50 nm for a typical situation. In the experiments, the angle θ at resonance can be related to the real part of ε_m, and the width of the resonance can be measured and related to the imaginary part of k_{zp}. To appreciate the sensitivity of the resonance to outside conditions, one can change ε_d slightly and observe the changes in the intensity of the reflected light. For example, we made a device using a BK7 glass prism and a silver film (figure 7.11). The reflecting surface was imaged on a camera and the resonance angle determined (by the way, the Scheimpflug construction (section 2.3) is relevant here!). Then a small jet of helium gas was played on to the exposed silver film and it was easy to visualize the flow of the gas because of the change in ε_d that the helium caused. Now the difference in refractive index between air and helium is only about 6×10^{-5}, so that the difference in ε_d is of order 10^{-4}; but of course the change observed is smaller than that because the gases are mixed in this experiment. This is typical of the sensitivity of SPR sensors, which are widely used as biological detectors where the dielectric used is water, and not air.

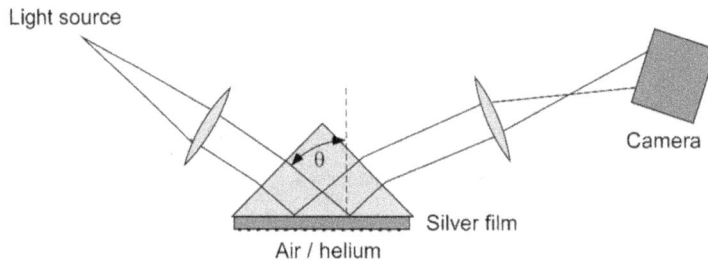

Figure 7.11. Setup for Kretschmann configuration with imaging camera.

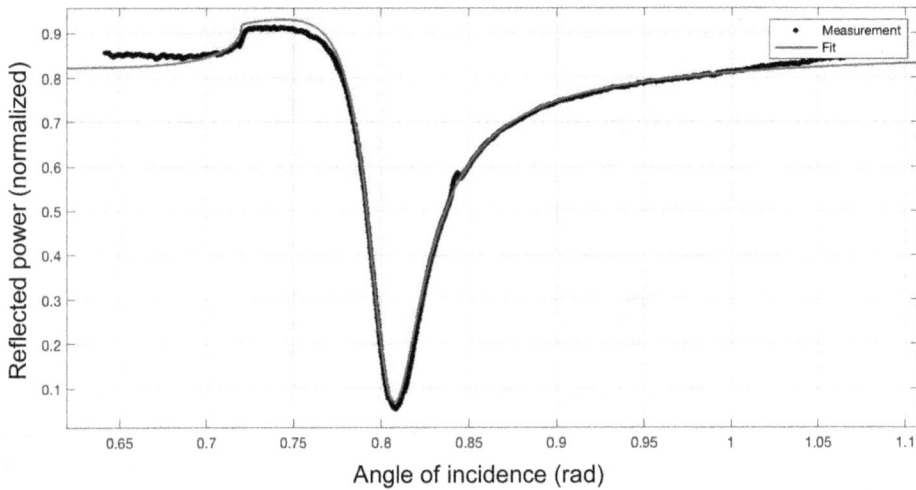

Figure 7.12. Results obtained for a 49 nm gold film at $\lambda = 633$ nm, fit to complex dielectric constant $\epsilon = -11.68 + 1.22i$ (blue curve).

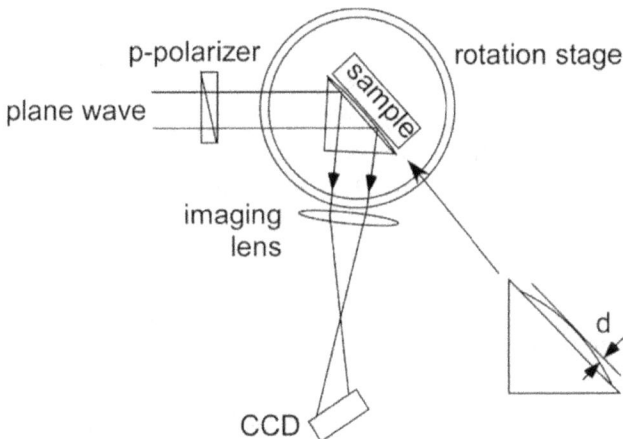

Figure 7.13. Setup for Otto configuration experiment, Bliokh *et al* [8].

7.5.3 Experiments using the Otto configuration

An experiment using the Otto configuration is discussed by Bliokh *et al* [8]. In this case, the glass hypotenuse has a convex shape, which is formed by optically contacting the plane surface of a long-focal length (1 m) plano-convex lens to the prism surface (figure 7.13). A planar metal surface touches the convex surface at one point, and the air thickness d at other points can then be calculated as a function of the radius r from that point. As in the Kretschmann configuration, the resonance occurs when the incident wave is at the angle where $k_z = k_0 n_d \sin \theta_{inc}$. Moreover, the amplitude of the reflected wave is a function of d; it is given by the usual multiple reflection equation (see section 3.3.4)

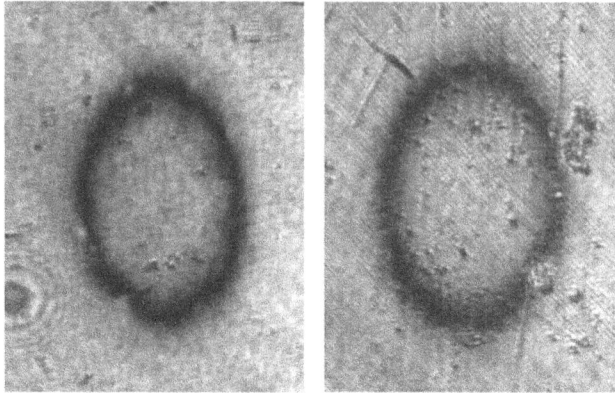

Figure 7.14. Resonance circles on a bulk copper plate (left) and on a copper film (right) on the same scale. Note that the circle is imaged as an ellipse, because of the need to use the Scheimpflug construction in imaging a plane inclined to the optical axis.

$$R(\theta) = \frac{R_{ga} + R_{am}\exp(-2d|k_y|)}{1 + R_{ga}R_{am}\exp(-2d|k_y|)}$$

where one notes that for the evanescent wave k_y is imaginary. In this equation, R_{ga} and R_{am} are the reflection coefficients calculated for the glass–air and air–metal interfaces, according to the usual Fresnel equations (section 3.2), taking into account that for incidence at an angle above the critical angle these coefficients are complex. What is observed in the experiment is that for particular values of d and θ, the value of $R(\theta)$ is minimum, essentially zero; for this geometry, the camera observes a circular resonance ring at radius r in the plane of the hypotenuse, which is imaged as an ellipse (figure 7.14), from which the value of d can be calculated. The numerical results relating these values of d and θ to the complex dielectric constant ε_m are shown in figure 2 of reference [8].

7.6 Induced optical anisotropy: photo-elastic, electro-optic and magneto-optic effects

7.6.1 Photoelastic effect

Many isotropic materials become birefringent when placed in fields of different types. A well-known example is the photoelastic effect, where a stress field causes an isotropic material to become uniaxially birefringent with axes locally parallel and perpendicular to the axis of the stress. The topic was first investigated and quantified by J C Maxwell in about 1850, using gelatine. The difference $n_o - n_e$ is proportional to the stress $\vec{\sigma}$. When the stressed material is observed between crossed polarizers, dark fringes occur at positions where $(n_o - n_e)d/\lambda$ is an integer, giving a set of 'isochromatic' fringes (so-called because their positions depend on λ), and also a set of 'isoclinic' fringes, which show where the stress axis is parallel to the polarizer axes, so that the polarization of the wave does not change. An example shown in figure 7.15 is a cantilever where the complete stress field is σ_x, and the polarizers

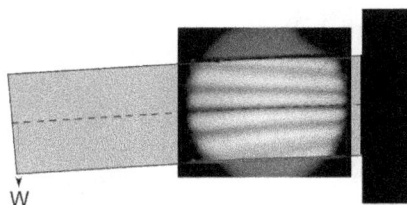

Figure 7.15. Isochromatic fringes in a cantilever beam made from Perspex (PMMA).

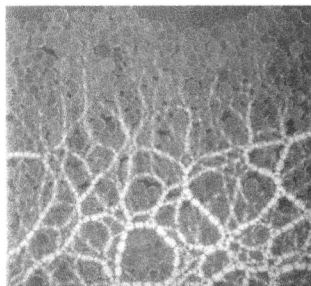

Figure 7.16. An array of discs of a photoelastic medium, in the space between two parallel glass plates, is observed between crossed polarizers. The discs have several diameters and are arranged randomly. Due to the weight of the discs alone, one can see the way in which contacts between them cause 'force chains'. Reproduced from [1] with permission of Cambridge University Press © 2011.

are at 45° to the x and y axes, so there are no isoclinics. The cantilever can be analyzed quantitatively, since the stress pattern is easily calculated; note that the pattern observed is independent of the sample thickness. Perspex was used as the medium; celluloid is probably better, but quite large forces are needed to get impressive results[2]. An extensive discussion of photo-elasticity is given in two volumes by Frocht [9]. Some special polymers with very high sensitivity have been developed to utilize the photoelastic effect; gelatine is an example of these. An example of investigation of force chains between discs with varying diameters in a two-dimensional array is shown in figure 7.16, in which the forces involved originate simply from the weight of the discs. In figure 7.17, we see a loaded Perspex beam, in which the stress is concentrated around the loading points. Figure 7.17(a) used polarizer and analyzer at ±45° (the isoclinics are at that angle) and in (b) they are at 0° and 90°. The isoclinics can be eliminated by the use of circularly polarized light.

7.6.2 Electro-optic effect

When an electric field is applied to certain transparent crystals they can be used to change the polarization state of a transmitted light wave. This electro-optic effect [10] has important applications such as fast optical switching. The linear effect, depending on $\vec{E}$ is called the 'Pockels effect', whereas at large fields a non-linear

---

[2] The photo-elastic effect can be investigated with much greater sensitivity using holography; see section 5.6.

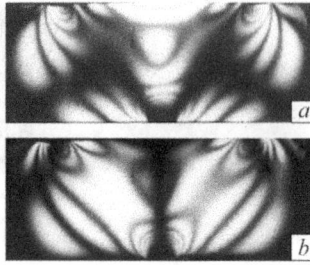

Figure 7.17. Isochromatics and isoclinics in a beam loaded at three points. The isochromatics are the same in both figures, but (a) the polarizers are crossed at 45° and the isoclinics (broad dark fringes) show regions where the principal axes are at ±45°, and in (b) the polarizers are crossed at 0°. Reproduced from [1] with permission of Cambridge University Press © 2011.

effect occurs, proportional to E^2 and called the 'Kerr effect'. The Pockels effect can be investigated conveniently in a crystal plate of KDP (KH_2PO_4), which has a tetragonal structure and is optically uniaxial with its optic axis in the 3-direction (z). It relates induced anisotropy in the (x, y) plane to an applied field E_3 in the z-direction, for example, so that a crystal which was uniaxial in zero applied field becomes biaxial.

A KDP crystal can be grown quite easily from aqueous solution.[3] A sample plate can be prepared by cutting from it a slice which is a few mm in thickness normal to the tetragonal axis (figure 7.18). Since the crystal is water-soluble, the plate can be cut and its surfaces polished using a wet disk, if other equipment is not available. For the experiment, thin (quite transparent) metallic electrodes need to be deposited on the two surfaces. Voltages up to about 1 kV are applied to the electrodes, and light propagates parallel to the electric field created. Since a voltage of this magnitude can be dangerous, the experimenters should be very careful. In particular, loose hair must be gathered up into a cap, so that it cannot accidentally touch the electrodes!

In terms of the dielectric-constant ellipsoid or indicatrix (section 7.2.1) of a tetragonal uniaxial crystal whose three principal semi-axes have values $\sqrt{\varepsilon_1} = n_o$, $\sqrt{\varepsilon_2} = n_o$, $\sqrt{\varepsilon_3} = n_e$, the Pockels effect changes the semi-axes $\sqrt{\varepsilon_1}$ and $\sqrt{\varepsilon_2}$ and creates a difference $\Delta n = AE_3$ between them proportional to the applied electric field E_3, causing the crystal to become biaxial. The total phase difference created between waves with polarizations 1 and 2 after propagation through the crystal, of thickness l, is then $\Delta\phi = k_0 l \Delta n = k_0 l A E_3 = k_0 A V$, where V is the applied voltage. The magnitude of the effect is usually defined by a 'quarter-wave voltage' $V_{\lambda/4} = \lambda/4A$, which has to be applied to create a quarter-wave path difference ($\Delta\phi = \pi/2$) between the principal waves, so that it behaves like a quarter-wave-plate (section 3.1).

In the experiment, a probe beam from an incident laser is initially polarized at 45° to the ellipsoid axes 1 and 2 which are in the plane of the plate. These axes have to be

---

[3] To get a good KDP crystal for this experiment requires a week or two of slow growth from saturated solution, so it should be started well in advance of the experiment.

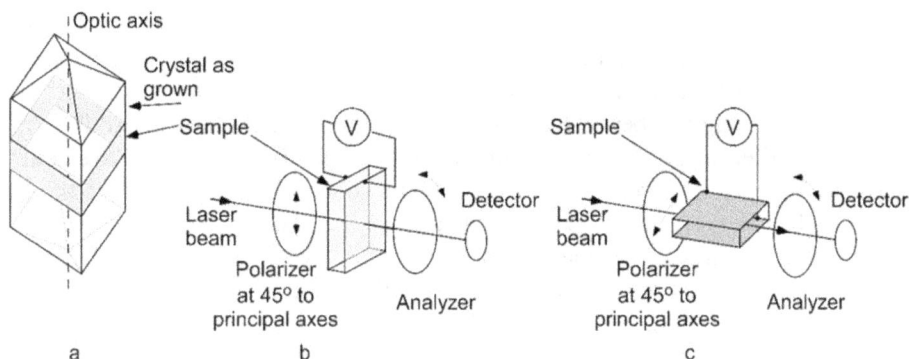

Figure 7.18. (a) KDP crystal and the Pockels effect experiment (b) longitudinal, (c) transverse.

identified, by finding the polarizations for which the voltage has no effect (the diagonals in the sample in figure 7.18(b))). The applied electric field creates the difference Δn between the two principal refractive indices in its plane and the wave becomes elliptically polarized. An analyzer polaroid which can be rotated in its own plane is used to measure the intensity ratio between the minor and major axes of the emerging elliptically-polarized wave as a function of applied voltage and the value of $V_{\lambda/4}$ can be determined, at which voltage the ratio is unity and the wave becomes circularly polarized. It is not necessary in the experiment to use a voltage source which reaches $V_{\lambda/4}$ (about 5kV); if accurate measurements of the ratio are made at smaller voltages they can be extrapolated since we know that while $\Delta\phi < \pi/2$, the ratio is $\tan^2\frac{1}{2}\Delta\phi = \tan^2 k_0 A V/2$.

The effect can also be seen in a transverse electric field; here the theory is a little more complicated [10] and there is a geometrical factor of length/ thickness, but smaller voltages may be needed and the electrodes can be made more substantial, since the light does not go through them. However the crystal will have to be cut differently so that two principal axes (1 and 3, or 2 and 3) lie in the plane of the incident wavefront.

7.6.3 Magneto-optic effect

The magneto-optic effect can be observed in a suitable homogeneous medium in a magnetic field. In this case, the characteristic waves are right- and left-handed circularly polarized waves, travelling parallel to the magnetic field. As a result, when a linearly polarized wave is incident on the sample, it exits with rotated axis of polarization (like in the corn-syrup experiments, section 7.3). Suitable easily-available materials are glass and water, although some more exotic materials show larger effects. The linear relationship between the angle of rotation and the magnetic field is denoted by the Verdet constant, which has values V(water) = 13 rad T^{-1} m^{-1} and V(flint glass) = 18 rad T^{-1} m^{-1}. As an example, the Verdet constant can be measured using a solenoid wrapped round a glass rod, providing say 10^{-2} T on a sample 0.1 m long. This provides a rotation of 18×10^{-3} rad = 1°. A laser beam is used to detect the effect with polarizer and analyzer. The two should be oriented relatively at 45° to get the largest sensitivity to changes in angle of rotation of the polarization of the wave (why?), but still the effect is rather small. Could it be increased by sending the beam back and forth several times along the rod?

Using an alternating current at mains frequency allows an improvement of sensitivity. When the current passes through the solenoid (remember that the field reverses each cycle) it results in periodic changes in the light intensity. These changes can be measured using a photodetector together with a magnetically-shielded phase-sensitive detector which reduces noise considerably, and a reasonably accurate value of V can be made.

References

[1] Lipson A, Lipson S G and Lipson H 2011 *Optical Physics* (Cambridge: Cambridge University Press)

[2] Berry M V, Bhandari R and Klein S 1999 Black plastic sandwiches demonstrating biaxial optical anisotropy *Eur. J. Phys.* **20** 1

[3] Powers P E and Haus J W 2017 *Fundamentals of Nonlinear Optics* 2nd edn (Boca Raton, FL: CRC Press)

[4] Yariv A 1987 *Quantum Electronics* 3rd edn (New York: Wiley)

[5] Maier S A 2007 *Plasmonics: Fundamentals and Applications* (New York: Springer)

[6] Simon H J 1975 Surface plasmons in silver films—a novel undergraduate experiment *Am. J. Phys.* **43** 630

[7] Pluchery O, Vayron R and Van K-M 2011 Laboratory experiments for exploring the surface plasmon resonance *Eur. J. Phys.* **32** 585

[8] Bliokh Y P, Vander R, Lipson S G and Felsteiner J 2006 *Appl. Phys. Lett.* **89** 021908

[9] Frocht M M 1941 *Photoelasticity* (New York: Wiley)

[10] Yariv A 1985 *Optical Electronics* (New York: Holt, Rinehart and Winston)

Optics Experiments and Demonstrations for Student
Laboratories

Stephen G Lipson

Chapter 8

Atmospheric optics

8.1 Rainbow: geometrical and physical optical effects, high-order rainbows

In natural surroundings, a rainbow is observed when the Sun is shining and it is also raining within the region of observation. For example, in the morning when the Sun is in the south-east, an observer looking towards a rainy region in the north-west will see the first-order rainbow at an angle of about 42° to the anti-Sun direction, which is conveniently relatively dark because of the cloudy background. Under good conditions, a second-order rainbow with reversed colour sequence at about 50° can also be seen, and the background light is noticeably darker in the space between the two rainbows (called 'Alexander's dark band'). Higher-order rainbows also exist, but the third and fourth orders would be in the direction of the Sun, and a fifth order, theoretically opposite to the Sun, is very weak and there are no reliable reports of its observation in nature. The rainbows are partially polarized in the tangential direction (figure 8.1). If the Sun is more than 42° above the horizon, the background is ground-cluttered, and it is difficult to see the rainbow; however, rainbows are often seen in waterfalls or sprays, where the complete circle around the anti-Sun direction might be visible.

8.1.1 The geometrical optical theory of the rainbow

The geometrical theory of the rainbow [1], deduced by Descartes in the 17th century, is well-known, and the colours are the result of dispersion of the refractive index of water. Since only angles are calculated, there is no dependence on the size of the water drops, provided that they are large compared to the wavelength (figure 8.2). However, there is a secondary effect consisting of what are called 'supernumerary bows', which are the result of interference and occur at angles depending on the ratio between wavelength and drop size. These bands appear adjacent to the

doi:10.1088/978-0-7503-2300-0ch8

blue end of the rainbow and are seen only when the raindrops have a small range of diameters.

Basically, the geometry of a refracted ray entering at offset b with respect to the centre of the drop, as shown in figure 8.3, can be described by the angles of incidence

Figure 8.1. Rainbow photographed through orthogonal polarizers.

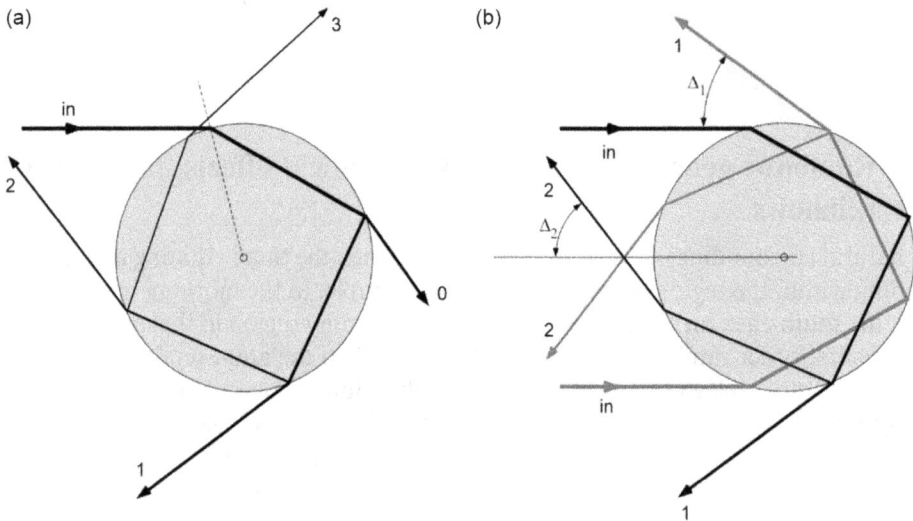

Figure 8.2. (a) The reflections of an incoming ray which gives rise to 1st, 2nd and 3rd order rainbows. The 0 order does not create a rainbow; it has no extreme angle of deviation. The 3rd order is seen in the direction of the Sun. (b) Note that when the whole sphere is illuminated, the 1st order reflection from an lower ray (red) exits at an angle Δ_1 close to Δ_2 for a 2nd order reflection from a parallel upper ray (black), so that the 1st and 2nd order rainbows are seen in proximity.

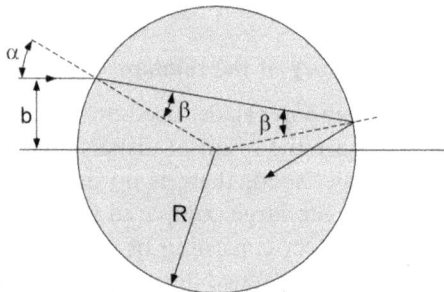

Figure 8.3. Refraction and reflection at the surface of a sphere.

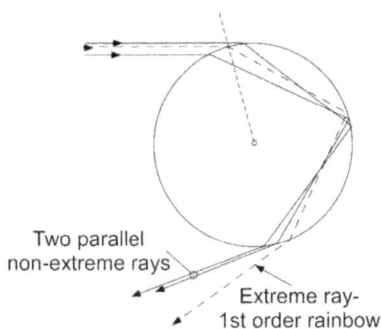

Figure 8.4. Rays which interfere to produce the supernumerary rainbows.

α and of refraction β, where $\sin \alpha = b/R$ and α and β are related by Snell's law and m is the number of reflections. It is found that the angle of deviation Δ of the exiting ray satisfies:

$$\Delta = 2\alpha - (2m + 2)\beta + m\pi.$$

At the extremum of order m we find that $\sin^2 \alpha = \frac{(m+1)^2 - n^2}{(m+1)^2 - 1}$, where n is the refractive index. To appreciate the variations in deviation as α is changed, it is a good idea to plot out Δ from the first equation as a function of α numerically, but the values of the extrema Δ_m of the various orders can be calculated from the above equations. It is important to appreciate that the function $\Delta(\alpha)$ is continuous, and so at the extremes of Δ the range of α contributing to the reflected beam $d\alpha/d\Delta$ diverges, and so the rainbow is quite intense. You see this clearly in the experiments.

The supernumerary bows (fringes parallel to the main rainbow) can be observed when the meteorological conditions are such that most of the raindrops have very similar diameters. The supernumerary bows arise from interference. Close to the extremum, there are two incidence angles α which give the same value of Δ, and therefore two waves which exit parallel to one another (figure 8.4). These waves interfere, constructively or destructively, as Δ changes. Obviously, the angles of these bows must be related to the dimensionless parameter λ/R, which is why the raindrops must mostly have similar diameters. The detailed calculation is given by Jackson [2].

8.1.2 Experiments

To investigate rainbow optics in the laboratory, we can explore several different directions.

1. A demonstration experiment used a spherical glass vessel filled with water and a white light source to show the dispersion effects. Does the fact that there is a container with a higher refractive index affect the situation significantly?
2. A second experiment used a cylindrical disc of transparent material to investigate the optics quantitatively in a two-dimensional analogue. Discs of reasonable quality can be cut from a Perspex ($n = 1.49$) cylinder about

50 mm diameter, while trying not to scratch the outer surface. Glass discs of higher quality with polished surfaces were also used. In this case the light source was a laser, and many orders could be found.

3. Using a small spherical ball of glass, illuminated by a laser, the basic geometry of the first order can be investigated, including the supernumerary bows. The surface of the ball needs to be well polished, and we have seen better results published, that were obtained with actual water drops.

We use a glass disc of radius R and a laser in order to observe rainbows of several orders (figure 8.5). The disc is mounted on a translation slide so that the laser beam can enter it at a variable position, the offset b from its centre lying between $-R$ and R. The laser should be at a position as close to the disc as possible, so that the beam is as narrow as possible. The various orders of reflection can then be identified (up to 7 have been observed). For each reflection the disc is translated and the exiting beam observed; it will be seen to move to an extreme angle, and then return. At the extreme angle, the beam is focused to a line, where it is brightest. We need to measure that extreme angle for each order. A direct measurement can be made as shown in the figure by intersecting the beam with a plane mirror, whose angle can be measured on a degree scale, and which is rotated until it returns the beam along its path. This can be seen by observing the point where it re-enters the disc, preferably slightly above or below the point where it came out, so that the two points can be easily aligned. The zero of the mirror's angle scale should be calibrated by replacing the disc by a retro-reflector (right-angle prism) before starting the experiment. To be able to measure angles in any direction, three rails (#2–4 in the figure) at 90° to one another are needed, as shown (why is a fourth rail not needed to complete the angular coverage?), but the carrier with the mirror and angle scale can be transferred from one rail to another as needed, and the angles supplemented by 90° and 180° as appropriate.

When all the angles of deviation Δ_m have been measured, the value of the refractive index of the disc can be determined. A criterion for a successful experiment is that all orders should give the same value of n, within the experimental accuracy.

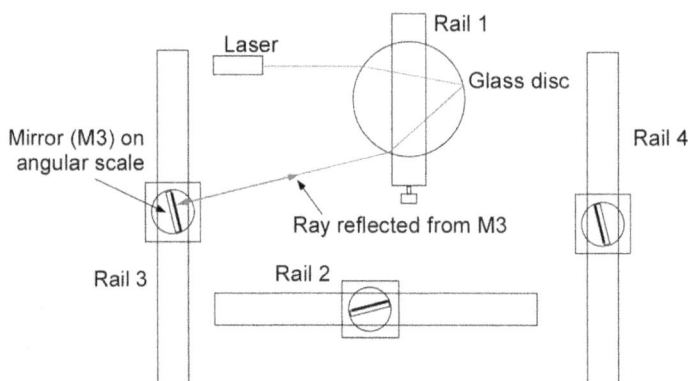

Figure 8.5. Experiment to measure the deviation angles of the rainbows of several orders.

An alternative method of doing the experiment is to use the mm scale on Rail 1 in figure 8.5 to determine b_m at the extrema, and hence α and Δ_m. This approach was found to be considerably less accurate because of extreme sensitivity to small errors in b_m, and gave different values for n for the various orders.

The polarization of the exiting laser beams (figure 8.1) can also be investigated. This arises from the Fresnel coefficients (section 3.2), and is most obvious when β is close to the Brewster angle.

Using back reflection from a ball lens, which is a small polished spherical glass ball, illuminated by a laser beam, it is simple to see that the light from a distant source is reflected into a cone (figure 8.6). This is the origin of the first order rainbow. If the refractive index of the glass is greater than 1.51, the second order is over the horizon. The monochromatic source and single spherical ball gives sharp supernumerary bands within the cone. The experiment in the figure used a 2.5 mm diameter ball lens made of SF10 glass ($n = 1.8$) and a He–Ne laser. The lens was stuck to a black card using double-sided transparent tape, and illuminated by a parallel laser beam. The reflected cone of light was projected onto a white card punctured by a hole at its centre to allow passage of the laser beam. The card was photographed by an external camera, and a magnified view of the edge of the reflected disc showed the supernumerary bows (figure 8.7).

8.2 Mirages and gradient-index optics

There are many situations in which light passes through a medium with spatially varying refractive index, so that the rays within the medium are not straight lines. One example is the graded-index optical fibre, which has maximum refractive index along its axis, and light is channelled along this axis even if the fibre is not straight. Examples in the atmosphere are mirages, resulting from light rays deviated by vertical temperature or pressure gradients in air, and observations of 'heat waves' in

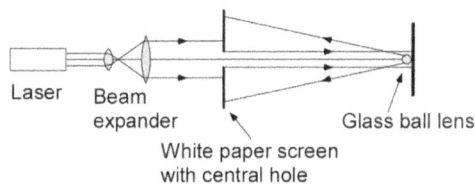

Figure 8.6. Experiment to show the reflected cone and the supernumerary bows, using a ball lens.

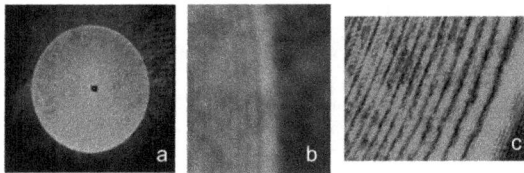

Figure 8.7. The light pattern reflected from a ball lens: (a) the reflected cone, (b) magnified region of (a) showing the supernumerary bows and (c) a contrast-enhanced rendering of the supernumerary bows.

light beams which are passing through turbulence where the temperature or pressure are spatially variant (see 'shadowgraph', section 4.4) [1, 3, 4].

8.2.1 Basic theory of ray paths

The analytic way of dealing with this problem is to use Huygens's principle in wave optics (using geometrical optics sometimes gives the wrong answer!) Let us look at a simple paraxial problem: a plane wave entering horizontally ($\theta = 0$) into a region where the refractive index $n(z)$ is a function of the height z only. At time $t = 0$, the wavefront is vertical (figure 8.8). Then, after a short time t, the wavefront at height z. will be at the value of x given by the optical path $n(z)x = ct$. For a given t, we then have a plane wavefront at angle $\theta = dx/dz$. From this, we can relate θ to the distance of propagation and we get a curved ray.

$$n(z)x = ct, \quad xdn + ndx = 0,$$
$$x(dn/dz)dz + ndx = 0,$$
$$\theta = -dx/dz = \frac{x}{n}\frac{dn}{dz}.$$

Finally, the radius of curvature of the ray is given by $R = \left(\frac{d\theta}{dx}\right)^{-1} = \left(\frac{1}{n}\frac{dn}{dz}\right)^{-1}$.

It is this curvature which gives rise to a mirage. The situation in a mirage seen over a hot road is that n of the air is smaller in the hot region next to the road and increases with height, as the air gets cooler. Thus dn/dz is positive, leading to rays which curve as shown in figure 8.9. Almost all the curve is next to the road, where dn/dz is largest, so that the ray behaves as if it is reflected from the road- and this gives the illusion that the road is wet. This is called an 'inferior mirage'. The angle of deviation is small, so the apparent reflection is only seen at almost grazing incidence. In northern climates, where in winter the ground is colder than the air above it, the opposite effect occurs. Then an ascending ray can be reflected by the upper warmer air and a superior mirage called 'fata Morgana' can occur; objects on the ground can be seen reflected in the sky!

The argument above can be used to show that an incident ray is reflected, i.e. the mirage is formed, under the same conditions as critical reflection. The incident ray at $z = 0$ and at angle θ_0 to the $\hat{x}$-axis is reflected from the plane at height z where

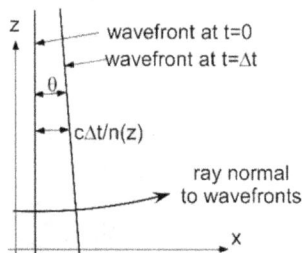

Figure 8.8. Curved ray due to an index gradient. © IOP Publishing. Reproduced with permission from [7].

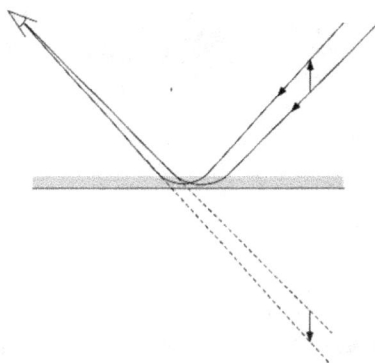

Figure 8.9. Formation of a mirage over a hot road.

$n(z) = n(0) \sin \theta_0$. See the appendix (section 8.2.3) for details. For example, if the air next to a road is at 50 °C, and the ambient is at 30 °C, we can see that the mirage is only observable at a glancing angle less than $\phi = \pi/2 - \theta$ to the road surface where $\phi^2/2 \approx \delta n/n_0 \approx 2 \times 10^{-5}$, $\phi \approx 6 \times 10^{-3}$ rad $= 0.3°$.

8.2.2 Laboratory experiments

Mirages: Experiments on inferior mirages can be done using a hot-plate to reproduce the outdoors situation, or on superior mirages using a tailored gradient of refractive index in water. In the former case, the hotplate has to be quite large, because the mirage is formed close to glancing angle. In addition, it is important to avoid specular reflection from the plate itself, which may be difficult.

The latter method is more amenable to quantitative experiments [3, 4], since $n(z)$ can be measured using a refractometer; in addition, situations are easily created where $n(z)$ is not linear, leading to focusing and other effects. The experiments can be done using an aquarium, in which we first put salt water to a depth of several cm, after which a layer of fresh water is carefully added. A convenient way to do this is to freeze fresh water in a tray (a photographic developing tray is ideal) and then float the ice block on the salt water. When the ice is floating, more water can be added if necessary by pouring it gently onto the ice. Look through the aquarium at a piece of lined paper and the refractive index structure which develops by diffusion is quite obvious (figure 8.10).

After a short while (say 30 min) a superior mirage can be seen by reflection in the interface between the salt and fresh water. What happens as time progresses is more complicated, because the refractive index profile developing is at first very non-linear, and only becomes linear after about 24 h. One quantitative experiment which can be carried out is to look at the curvature of a laser beam passing through the water (figure 8.11).

At the same time, the refractive index profile can be measured by carefully extracting with a syringe samples of the water as a function of depth, and measuring their indices with an Abbe refractometer (section 2.2). The observed beam profile is digitized, and the curvature calculated at each point; this can then be compared with

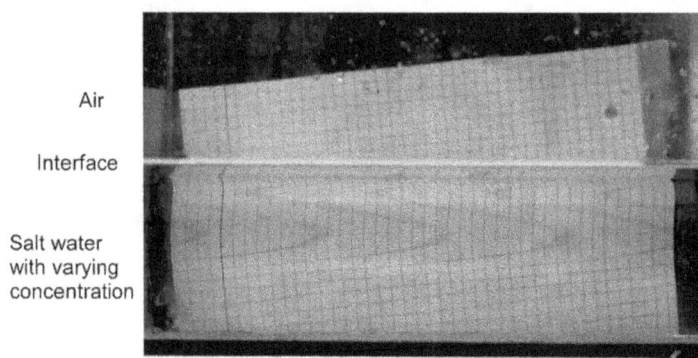

Figure 8.10. Qualitative observation of index gradients developing in salt water. © IOP Publishing. Reproduced with permission from [7]. All rights reserved.

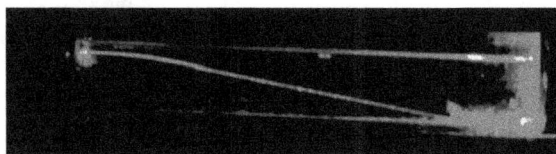

Figure 8.11. A laser beam enters the index gradient region horizontally from the left. Note that the lower two-thirds of the medium has negligible index gradient; the light ray is a straight line. © IOP Publishing. Reproduced with permission from [7]. All rights reserved.

the expected form $\frac{1}{R} = \frac{d^2y}{dx^2} = \frac{1}{n}\frac{dn}{dz}$. Note that this is only true for small deviations from the horizontal; the approach used in the appendix (section 8.2.3) can be developed for larger deviations. The propagation of the light forming a superior mirage can also be followed using a laser in this experiment, when the laser in figure 8.11 enters with an upward gradient at a suitable height.

Gradient-index lenses: A gradient-index (GRIN) lens is basically a thin plate of thickness d in the x–y plane with refractive index $n(r) = n_0 - br^2$. Then by using Fermat's theorem it is easy to see (to a quadratic approximation in r) that this acts as a lens with focal length $1/2bd$. Unfortunately, commercial GRIN lenses available have very small diameters (<2 mm) since they are mainly used for fibre-optics applications, so they are not suitable for a lab experiment.

8.2.3 Appendix

Consider a ray at angle θ to the x-axis (figure 8.12). The curvature of the ray, when it is travelling along the x-axis, was seen to be $\frac{1}{R} = \frac{d\theta}{dx} = \frac{1}{n}\frac{dn}{dz}$; now, when it is travelling at angle θ to the x-axis, it is determined by the projection of $\frac{dn}{dz}$ on the wavefront which is in the plane normal to the ray, i.e. $\frac{1}{n}\frac{dn}{dz}\cos\theta$. In general, we can use distance s and θ to measure the position and wavefront angle at a point along the ray, and then $dz = ds\sin\theta$. The curvature is

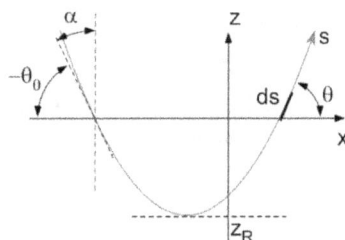

Figure 8.12. Geometry of a curved ray in a stratified medium.

$$\frac{1}{R} = \frac{d\theta}{ds} = \frac{d\theta}{dz} \sin\theta = -\frac{1}{n}\frac{dn}{dz} \cos\theta.$$

The equality between the last two expressions in the equation above can be re-written as

$$\frac{\sin\theta . d\theta}{\cos\theta} = \frac{-d(\cos\theta)}{\cos\theta} = \frac{dn}{n}.$$

We can integrate this from a starting angle $\theta_0 = -(\pi/2 - \alpha)$ relative to the x-axis at $z = 0$ (figure 8.12), to grazing angle $\theta = 0$, where α is the angle of incidence relative to the $-z$-axis, which is the usual formulation, in order to find the refractive index $n(z_R)$ at the height z_R where the ray is reflected:

$$-\int_{\theta_0}^0 \frac{d(\cos\theta)}{\cos\theta} = \int_{z=0}^{z_R} \frac{dn}{n}$$

$$-\left[\ln(\cos\theta)\right]_{\theta_0}^0 = \left[\ln n\right]_{z=0}^{z_R}$$

$$\ln(\cos\theta_0) = \ln(n(z_R)/n(0))$$

$$\cos\theta_0 = \sin\alpha = n(z_R)/n(0)$$

$$n(0)\sin\alpha = n(z_R)$$

which is the critical reflection condition.

8.3 Green flash

The 'green flash' is an optical phenomenon (*not* an illusion!) which can occasionally be observed at sunset over a marine horizon at the last moment as the Sun disappears below the horizon. For a moment, a green or turquoise flash is seen. It lasts for a fraction of a second in temperate latitudes, but can last for seconds or minutes closer to the Earth's poles, where sunset is a slower phenomenon. It can also be seen at sunrise, when the green flash precedes the sunrise. For a long time, it was thought to be an after-image on the retina of the observer who had, after all, been looking at the red Sun a moment earlier! Only when the phenomenon had been photographed on film, and observed at sunrise, was it accepted as a physical observation which required an explanation [5, 6].

8.3.1 Physical origin of the green flash

The explanation of the green flash depends on the gradient of refractive index of the air above the Earth's surface, resulting mainly from the atmospheric temperature profile. The pressure reduction with height has negligible influence. As a result, the atmosphere behaves as a thin prism with its base on the horizon. When we see the Sun setting, it is actually about 0.6° below the observed position. In addition, the prism causes dispersion (the Abbe dispersion index of air is about 90), and the last bit of the Sun's image to disappear below the horizon at sunset is in the blue end of the spectrum. However, almost all of the blue light from the Sun is scattered out by Rayleigh scattering (section 3.4) during the long path of the Sunlight through the atmosphere at sunset (about 350 km, compared with 9 km when the Sun is at its zenith), which is why the setting Sun is red. So the last bit of the Sun's rim to disappear is green (figure 8.13). The angular dispersion is 0.6° divided by the Abbe dispersion index (90), which is about 0.006°, less than the resolution of the naked eye (a perfect eye would have to have an iris diameter of more than 5 mm to attain this resolution), so that most photographs of the green flash have been taken with long-focal-length cameras or through telescopes.

However, the fact that the green flash is occasionally seen by casual observers without a telescope indicates that some more physics is involved. Most casually-observed green flashes are the result of an inversion layer in the atmosphere, which is a layer of cool air located just above the ground, above which the temperature rises again before falling at higher elevations. Such inversion layers often trap pollution, but this is less noticeable over the sea. The resulting refractive index profile then acts as a GRIN lens and concentrates the residual sunlight in the direction of the observer, and the focus has chromatic aberration, which is the origin of the observable green flash.

8.3.2 A laboratory experiment

An experiment to illustrate this phenomenon was described by ben Aroush *et al* [7] in which salt-water with height-dependent concentration and refractive index obtained by diffusion was used to simulate the atmospheric profile. The scaling of

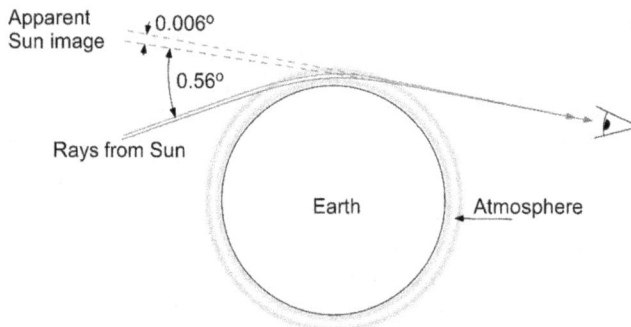

Figure 8.13. Light rays from the Sun refracted and dispersed through the Earth's atmosphere. © IOP Publishing. Reproduced with permission from [7]. All rights reserved.

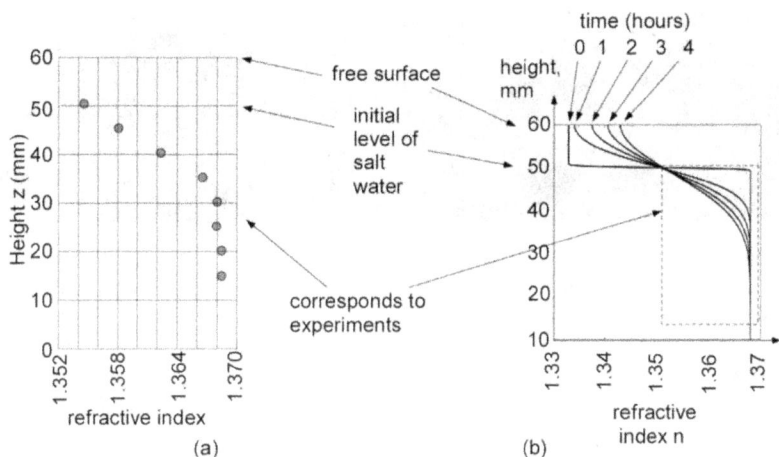

Figure 8.14. Vertical refractive index profiles (a) measured and (b) calculated for the salt-water tank.

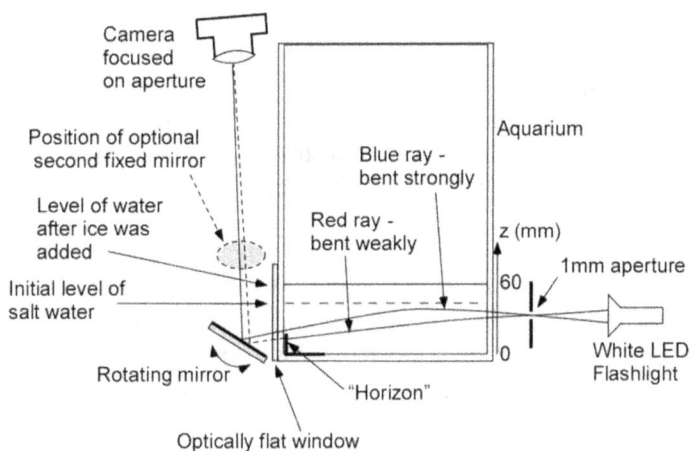

Figure 8.15. Experimental arrangement used to observe the 'blue flash'.

refractive index from air to water results in similar conditions to the atmospheric situation over a distance of 350 km being obtained in about 25 cm of water. The basic idea is the same as that used for observing mirages in section 8.2. A calculation of the refractive index profile obtained by diffusion is shown in figure 8.14, showing that suitable conditions are obtained after a few hours of diffusion of fresh water into salt water. The experimental arrangement is shown in figure 8.15. In fact, when you look into the light path and lower or raise your head (corresponding to sunset and sunrise) you see the refracted image of the source and a mirage of it, and they merge as your head moves, creating the green flash. Figure 8.16 is a sequence of photographs showing this, which well corresponds with the natural phenomenon. The sequence is also available as a movie. Actually, in the lab it is rare to find a

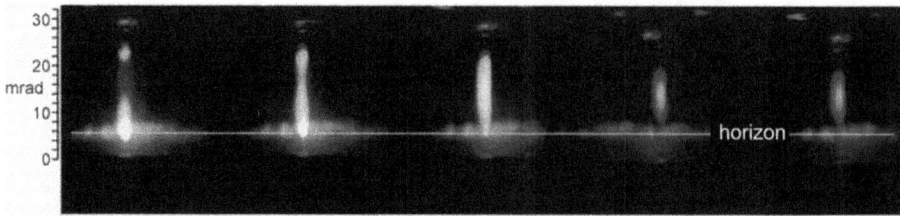

Figure 8.16. Experimental results showing the merger of the superior mirage with the direct image of the source to create the blue flash. © IOP Publishing. Reproduced with permission from [7]. All rights reserved. A movie is available online at http://iopscience.iop.org/book/978-0-7503-2300-0.

white-light source which doesn't contain enough blue light to create a blue flash rather than a green one (we tried using a filament lamp run at low voltage, but even that created a blue flash!), but the idea is valid. As in the mirage experiments, the refractive index profile can be measured with an Abbe refractometer.

8.4 Sky polarization, the sunstone and Viking navigation

8.4.1 How the Vikings used a birefringent crystal for navigation

Before the invention of the magnetic compass, the Vikings sailed to many destinations from America to western Europe. The question of how they navigated, often under bad weather conditions, is of historical interest. One suggestion, which seems to have archaeological backing[1], is that they used the polarization of sky-light to identify the Sun's position in the sky, by the help of a birefringent crystal [8, 9].

If we take a birefringent crystal such as Calcite, also known (significantly?) as Iceland Spar, and look at an object through it, we generally see two images, having orthogonal polarizations called 'ordinary' and 'extraordinary' (figures 8.17 and 3.1(c)). If the object is scattering unpolarized light towards us, the two images have equal intensity. However, if the scattered light is partially polarized, the images have differing intensities, which depend on the relative orientation of the polarization direction and the principal axes of the crystal dielectric tensor. Now when sunlight is Rayleigh-scattered by the air, as we saw in section 3.4, it becomes polarized with the electric vector of the scattered light normal to the plane defined by the observer, the Sun and the scattering region. If the scattering is weak and the scattering angle is 90°, the degree of polarization can be quite high (see figure 3.11), but even if the scattering is stronger, and multiple scattering becomes significant, the partial polarization is still predictable and is strong enough to be detected. The use of polarization to determine the Sun's position in the sky is based on this idea.

At first, we might consider using a polarizing film to do this. Choose a point in the sky and observe it through the polarizer. Simply rotate it until the transmitted intensity is lowest, and we have the direction of the polarization of the scattered light. However, when the sky's degree of polarization is weak it is very difficult to

---

[1] Calcite crystals were found in the remains of shipwrecked boats, and were recently identified as sunstones [8], and 'sunstones' were mentioned in Icelandic sagas from the 10th century.

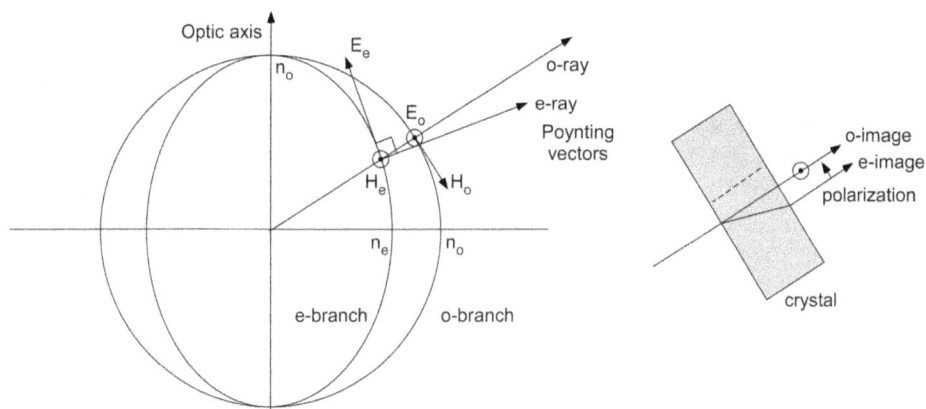

Figure 8.17. The refractive index surface of calcite ($n_o = 1.66$, $n_e = 1.49$) and the formation of two images when unpolarized light passes through an arbitrarily oriented crystal. Note that the o-image is polarized normal to the plane containing the two images, and the e-image is polarized parallel to it. When the two images are equal in intensity, the incident light is polarized at either ±45° from this plane.

determine the angle of minimum intensity accurately enough for this to be useful. Using a birefringent crystal allows us to do this by comparing the intensities of the ordinary and extraordinary images of an aperture illuminated by the sky; this is the basis of the sunstone. As the crystal is rotated, the relative intensities of the two images change, and when the two are equal, the vector joining the two images is at 45° to the electric vector of the polarized light. Because this is a comparison between two intensities of identical apertures, which can be situated in close proximity, this angle can be determined very accurately by eye even if the degree of polarization is weak. However, there are two possible polarizations (at ±45°) for the incident light and supplementary information is needed to distinguish between them; this might be provided by turning the crystal 45° and finding which image is brighter, or by comparing two observations made on different patches of blue sky. The aperture can be replaced by a black spot painted on the crystal (this seems to be how the Viking sunstone was made), and the same method used, since each image of the black spot obscures one of the polarizations. What is important is that the size of the aperture or spot be calculated so that the ordinary and extraordinary images just touch one another, so that the comparison is most accurate. Lab experiments indicate that an accuracy of about 1° in the rotation angle can be achieved with a degree of polarization of 0.1 [10]. What is also significant from the point of view of navigation is that this determination of the Sun's position can be carried out when the Sun is below the horizon, provided that there is enough light in the sky to make the observation.

8.4.2 A Sunstone in the laboratory and the open air

A laboratory experiment can be made to confirm this idea by providing a screen illuminated with partially polarized white light (unpolarized light plus an additional polarized component) and using the crystal to determine the angle of the polarized component. It is important to assess the accuracy of the method, as a function of the degree of polarization.

The sunstone can be used to investigate the sky polarization as follows. When multiple Rayleigh scattering is taken into consideration, the picture of the sky polarization is more complicated [11]. In figures 2 and 3 of reference [11], stereoscopic pictures of the polarization distribution are shown for single and double scattering, the latter of which corresponds well to experimental investigations. What is striking is that there are four singular points in the sky where the degree of polarization is zero, i.e. not polarized, and their positions are 10°–20° above or below the Sun and anti-Sun positions. The sunstone is an ideal device for finding these points. Three of them are known by the names of their discoverers—the Brewster point (1846) above the Sun and the Babinet point below the Sun (1840), and the Arago point (1809), which can be observed just after sunset, above the anti-Sun position. The fourth point, under the anti-Sun position, is not observable from the Earth's surface, but has been located from balloon or plane flights. To make a good determination, the Sunstone device can be mounted at the end of a black cardboard tube, which can then be oriented more accurately at a particular orientation using a telescope tripod. The tube and the Sunstone can be rotated together, or the Sunstone rotated independently, to discover the neutral points where the sky is unpolarized. With some effort, the direction of polarization of other points in the sky can also be found and compared with figures 5 and 7 in [11]. A fuller investigation of the sky polarization on a clear day could be carried out photographically by taking several pictures in the same direction, with at least three different orientations of a polarizer in front of the lens [12].

References

[1] Greenler R 1991 *Rainbows, Halos and Glories* (Cambridge: Cambridge University Press)

[2] Jackson J D 1999 From Alexander of Aphrodisias to Young and Airy *Phys. Rep.* **320** 27–36

[3] Vollmer M and Greenler R 2003 Halo and mirage demonstrations in atmospheric optics *Appl. Opt.* **42** 398

[4] Vollmer M and Tammer R 1998 Laboratory experiments in atmospheric optics *Appl. Opt.* **37** 1557

[5] Young A T 1999 Green flashes *Opt. Photonics News* **10** 31–7

[6] Young A T *A Green Flash Homepage* http://aty.sdsu.edu/index.html

[7] ben Aroush T, Boulahjar S and Lipson S G 2018 Observing the green flash in the laboratory *Eur. J. Phys.* **39** 015301

[8] Horváth G, Barta A and Pomozi I *et al* 2011 On the trail of Vikings with polarized skylight: experimental study of the atmospheric optical prerequisites allowing polarimetric navigation by Viking seafarers *Philos. Trans. R. Soc. B* **366** 772–82

[9] Bernáth B, Blahó M and Egri Á *et al* 2013 Orientation with a Viking sun-compass, a shadow-stick, and two calcite sunstones under various weather conditions *Appl. Opt.* **52** 6185

[10] Ropars G, Gorre G and Le Floch A *et al* 2012 A depolarizer as a possible precise sunstone for Viking navigation by polarized skylight *Proc. R. Soc. A* **468** 671–84

[11] Berry M V, Dennis M R and Lee R L Jr 2004 Polarization singularities in the clear sky *New J. Phys.* **6** 162

[12] Lee R L Jr 1998 Digital imaging of clear-sky polarization *Appl. Opt.* **37** 1465

IOP Publishing

Optics Experiments and Demonstrations for Student Laboratories

Stephen G Lipson

Chapter 9

Relativistic optics

9.1 Fizeau's experiment: velocity of light in moving water

In 1851, more than a century and a half ago, Fizeau [1] carried out an experiment to measure the velocity of light in a moving medium and obtained results which could only be explained, some 50 years later, using the special theory of relativity. It is one of the few experiments in the field of relativistic optics in which the measured effect is linear in the velocity ratio v/c, as opposed to quadratic, and could therefore be done with relatively simple apparatus. The experiment compared the velocity of light moving upstream and downstream in moving water. The velocity of light in the stationary medium is c/n, where n is the refractive index.

According to classical dynamics, if the medium moves at velocity v in the same direction as the wave, the wave would propagate at velocity $c_+ = c/n + v$ in the lab frame, and at $c_- = c/n - v$ if the motion is in the opposite direction, giving effective refractive indices $n_\pm = n/(1 \pm vn/c) \approx n(1 \mp vn/c + ...)$ in the two cases. From which,

$$\delta n = n_- - n_+ = 2n^2 v/c$$

is the difference between the two indices. If coherent light waves travel distance L in the lab frame in both directions relative to the moving medium simultaneously, and the waves interfere after leaving the medium, a phase difference $\delta\phi = 2\pi L \delta n/\lambda = 4\pi L n^2 v/c\lambda$ would occur. Surprisingly, even if the medium is a vacuum for which $n=1$, a phase difference would be observed! Something must be wrong.

According to the theory of special relativity, $c_\pm = (c/n \pm v)/(1 \pm v/nc)$, from which

$$\delta n = 2n^2 v(1-1/n^2)/c.$$

The phase difference becomes $\delta\phi = 4\pi L n^2 v(1-1/n^2)/c\lambda$. Now, when $n = 1$, there is no effect. When Fizeau's experiments showed the latter expression to fit his data, the factor $(1-1/n^2)$ was identified as the 'aether drag coefficient' which had already been discovered in astronomical aberration experiments by Fresnel, but whose origin was unknown at the time.

Fizeau's apparatus was based on the Rayleigh interferometer, which is a Young's slit experiment in which the phase difference between light passing the two slits is modified by a medium to be measured, and is shown in figure 9.1. His idea was improved by Michelson [2], who used the idea of the Sagnac interferometer (section 5.3), which automatically increases the sensitivity by a factor of 2. But of course neither of these experimentalists had lasers available, which must have made the experiments very difficult, and emphasizes their incredible ingenuity and patience! In figure 9.2, we show such a Sagnac system which circulates a laser beam around a common path in two senses, and the interference pattern between

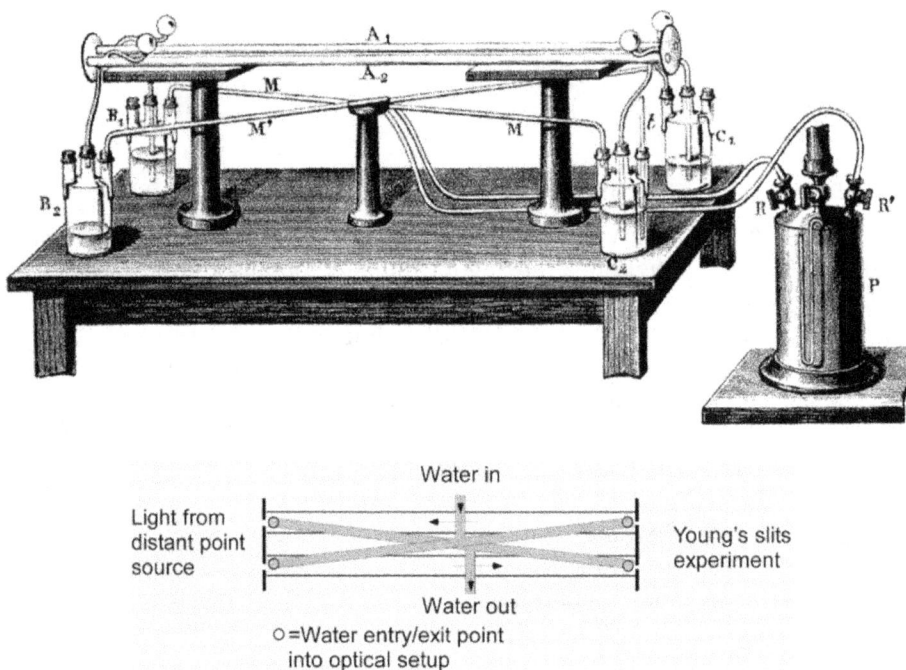

Figure 9.1. Fizeau's [1] experimental system. This is fairly stable in the presence of mechanical disturbances. Reproduced from https://en.wikipedia.org/wiki/Fizeau_experiment#/media/File:Fizeau-Mascart2.png.

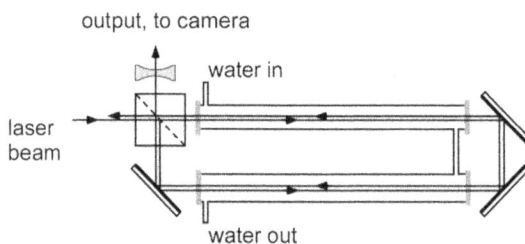

Figure 9.2. Sagnac interferometer system.

them is very stable since most disturbances, such as mechanical movements of the optical elements or air currents, affect both beams identically. In this case, one light wave goes in the same direction as the water flow, and the other one in the opposite direction.

Let us estimate the size of the effect for water ($n = 1.33$, so that the drag coefficient is about 7/16) flowing in a tube with a total length of $L = 2$ m, at a velocity 1 m s^{-1}. Then $\delta\phi = 4\pi L n^2 v(1-1/n^2)/c\lambda \approx 10^{-1}$ rad, which is a fraction of an interference fringe. How can we make it bigger? One way is to increase the size of the apparatus and/or the velocity of the water flow; the figures given above are quite modest, and Michelson, when he repeated Fizeau's experiments in about 1890, used a path of 3 m and a velocity of 8 m s^{-1}, which leads to nearly one fringe, compared to the classical estimate of about 3, as described in his book *Studies in Optics* [2]. Another way is to circulate the light several times through the water although, as Michelson points out, it is important to ensure that the beams sample water moving at a known velocity, which might not be constant across the diameter of the tubes used. On the other hand, because the Sagnac interferometer is very stable, it is possible to detect very small movements in the interference fringes using a fast camera and image processing, which were not available to Michelson, and this is probably the best way to get good results today.

To measure the velocity of the water flow, it is only necessary to collect the water and measure the volume flow in a given time. This volume has then to be divided by the cross-sectional area of the tubes surrounding the laser beam. However, the flow vector of the water around the entrances and exits must be considered, because it might not be exactly parallel to the laser beam in these regions. Michelson used symmetrical plumbing to ease this problem.

9.2 Optical fibre gyroscope: measurement of rate of rotation

9.2.1 Sagnac interferometer in a rotating frame of reference: optical gyroscope

A unique application of the Sagnac interferometer (section 5.3) is the optical gyroscope [3]. If the whole interferometer (either version) is built on a table which can be rotated at angular velocity Ω, the interference fringes shift by phase $8\pi\Omega A/c\lambda$, where A is the total area enclosed by the light path [4]. This is called the 'Sagnac effect'. The basic idea can easily be proved for a model circular loop of radius r. If the light takes time T to traverse the interferometer from the beam-splitter back to the beam-splitter, this element has moved a distance $Tr\Omega$, which represents a $2Tr\Omega/\lambda$ fringe shift, since for one sense of rotation the path is increased and for the other it is decreased. Now the time T, for a circular loop, has value $2\pi r/c$ so that the fringe shift is $4\pi r^2\Omega/c\lambda = 4A\Omega/c\lambda$, and the phase shift is 2π times this. Schwartz and Meitav [4] show that this is generally true, even for a non-circular path for which $A \neq \pi r^2$, and is also independent of the position of the axis of rotation, even if it is outside the loop. This result, derived from Newtonian mechanics, is also true in special relativity. If you put in some numbers, the effect can be seen to be measurable without too much trouble. For an area $A = 1$ m^2, and a rotation rate of 1 rotation per sec, we have a 0.2 fringe movement, which can easily be observed. Optical

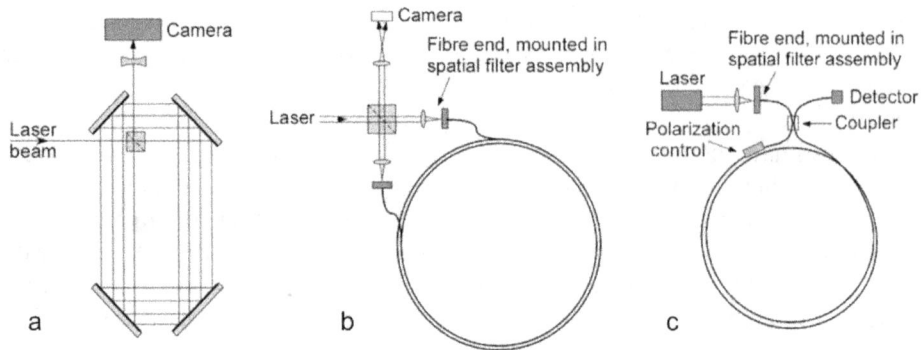

Figure 9.3. Sagnac interferometers on a rotating optical table: (a) rectangular light path with several turns, (b) using a coiled fibre and a beamsplitter and (c) using a coiled single-mode fibre and a coupler.

gyroscopes use an optical fibre to increase A by orders of magnitude, so that very small values of Ω can be measured accurately. When the waves travel in a medium where the refractive index n is not unity, Newtonian mechanics add a factor n^2 to the fringe movement, but this disappears when the calculation is done with special relativity (prove this!). The Sagnac effect experiment can be done either using an up-graded version of figure 5.13, where the beams make several turns before interfering, so as to increase A (figure 9.3(a)), or using fibres.

9.2.2 Fibre-optical gyroscope

A practical gyroscope (figure 9.3(b)) uses a long optical fibre. In this system, the optics for focusing the laser beam into and out of the fibre ends can be constructed by modifying a laser spatial filter setup, in which the fibre replaces the pinhole. In figure 9.3(c), the ends of the fibre are connected to the two outputs of a commercial fibre coupler, which is the fibre equivalent of a beam-splitter. A polarization controller, essentially a controlled twister, has been added to improve contrast in the output. These versions are described in detail by Schwartz [5]. In both of these systems, the laser and camera have to be battery-operated, so that they can be fixed to the rotating table. Moreover, the camera has to be wirelessly operated, or operated single-frame with enough delay to allow the table to come to a constant rotation rate before the fringes are photographed. Alternatively, a video camera can be used; in this case, a fixed object can be arranged to interrupt the laser beam once in every cycle so as to record the rotation rate as part of the video. When the fibre used is a single-mode fibre, the output is not an interference pattern with fringes, but a Gaussian field with intensity which changes with the rate of rotation. The light output is divided between the two exits (one back into the laser and one in the direction shown) so that energy is conserved. To make this quantitative, it has to be calibrated, since the intensity varies sinusoidally as a function of the angular velocity, between maximum and minimum values which must be determined.

References

[1] Fizeau H 1851 Sur les hypothèses relatives à l'éther lumineux *C. R.* **33** 349–55
   English: Fizeau H 1851 The hypotheses relating to the luminous aether, and an experiment which appears to demonstrate that the motion of bodies alters the velocity with which light propagates itself in their interior *Philos. Mag.* **2** 568–73

[2] Michelson A A 1927 *Studies in Optics* (Chicago, IL: University of Chicago Press) reprinted by Dover, 1995

[3] Lefèvre H 1993 *The Fiber-Optic Gyroscope* (Boston: Artech House)

[4] Schwartz E and Meitav N 2013 The Sagnac effect: interference in a rotating frame of reference *Phys. Educ.* **48** 203

[5] Schwartz E 2017 Sagnac effect in an off-center rotating ring frame of reference *Eur. J. Phys.* **38** 015301

IOP Publishing

Optics Experiments and Demonstrations for Student Laboratories

Stephen G Lipson

Chapter 10

Basic experiments in quantum optics

The concept of a 'photon', which represents the quantum of energy of an electromagnetic wave, is nowadays taken for granted. However, it is important to establish experimentally the validity of the concept, and to confirm its value as the Planck constant h times the wave frequency ν.

It is often stated that the photo-electric effect, in which an electromagnetic wave impinges on a photo-cathode surface and causes the emission of electrons, proves the existence of photons. The photo-emission occurs provided that the value of $h\nu$ exceeds the work function, which is the energy with which the electron is bound to the cathode, and the kinetic energy of the emitted electron is then equal to the difference between these two energies. This seems to establish the existence of a quantized state of the electromagnetic field, but in fact can be explained using a semi-classical theory in which a classical electromagnetic field perturbs an electron in a quantized system (the photocathode) which is excited from its ground state (inside the cathode) to an excited state in which it is free and has kinetic energy [1]. In fact, the proof of existence of photons as quantized entities requires more sophisticated techniques employing correlators.

10.1 Coincidence experiments

To make the point, suppose that a weak light wave is incident on a beam-splitter, at an average rate of one unit of energy $h\nu$ per time T. If the wave is described as a stream of photons, then each photon must leave the beam-splitter by one of its two ports, either the reflecting or transmitting one. It cannot leave by both of them. So a decisive experiment (figure 10.1(a)) would seem to be to detect emissions from both ports (using ideal detectors D1 and D2 each of which ejects an electron in response to the absorption of a photon) and look for coincidences within time T. There should be none. If the same experiment is described classically, the wave will exit with half

doi:10.1088/978-0-7503-2300-0ch10

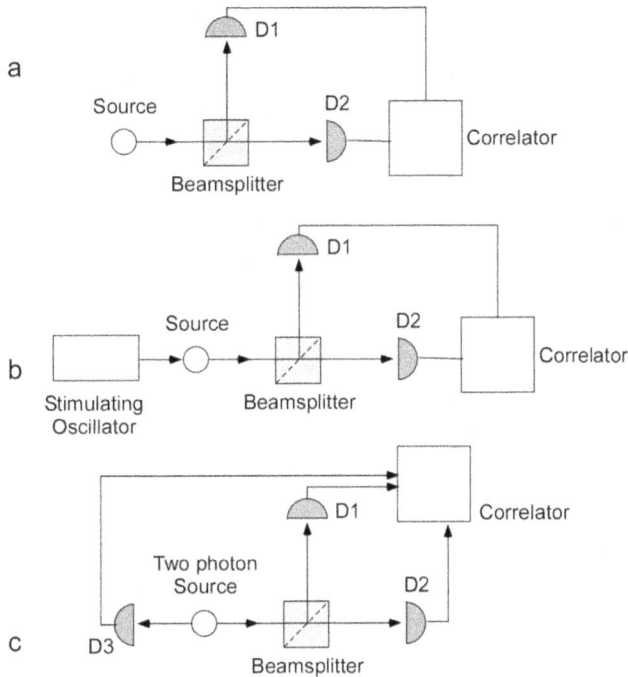

Figure 10.1. Three experiments to detect coincidences. (a) The source emits photons randomly, (b) the source is stimulated to emit photons by an external periodic oscillator [4] and (c) the source emits pairs of correlated photons [2, 3].

the power from each port, and when the wave reaches each detector it acts to perturb its cathode and will have 50% probability of causing ejection of an electron from each of them within time T. This does not preclude the possibility of both of them ejecting, which would have a probability of 25%. Likewise, there is a 25% probability that neither will eject an electron. So it would seem that the classical and quantum descriptions can be distinguished by looking for coincidences, which will be absent in the quantum case.

However, if the light source emits photons at an average rate of one per time T, and the distribution of emission times is *statistically random*, i.e. a Poisson distribution, there is a non-zero probability that two or more photons will be emitted within the time T, or conversely that no photon will be emitted. If two photons are emitted, there is a probability that each one will exit the beam-splitter by a different port, so that a coincidence is recorded within T. The probability of this happening is about 25%, so that even if the photon picture is valid, the number of coincidences is not zero, but has a similar value to the classical case.

A more exact calculation for the Poisson case shows indeed that quantum and classical calculations coincide. For the Poisson distribution with an average rate of s events in a given time interval, the probability of m events being measured (assuming unit quantum efficiency of detection) is $P(m) = e^{-s}s^m/m!$. As a result, if $s = 1$ (which defines the time interval T), the probability of $m = 1$ emission is $P(1) = e^{-1}$, while that of no emissions is also $P(0) = e^{-1}$ and that of two emissions within time T is $P(2) = e^{-1}/2$, which is not negligible. Then, the probability that one emission is reflected and one transmitted by the beam-splitter is 1/2, so that the probability of a 'coincidence' is $\frac{1}{2}P(2) = e^{-1}/4$, which is 25% of that for a single emission in either channel [1].

So what is needed is *a light source which is not Poisson*, which is called 'squeezed' or 'anti-bunched'. The ideal squeezed source emits photons at regular intervals T, (figure 10.1(b)) and then the existence of photons can be demonstrated by anti-coincidence, since the possibility of two photons being emitted within the coincidence window has been avoided. Quantum wells are such sources and have been used to show that indeed coincidences do not exist within time T. Another variant on the theme is to use sources which emit two photons coincidentally [2, 3] (figure 10.1(c)) in which case one photon goes to the beam-splitter and the other one goes to a third detector D3. One then expects that there will be coincidences between D3 and *either* D1 *or* D2, but never all three. These experimental set-ups [4] require more sophisticated equipment than other experiments in this book, but the experiment is very important and that shown in figure 10.1(b) was carried out in a neighbouring laboratory to mine. Today, experiments on photon correlations for student labs can be purchased, but I have not yet had the chance to use such systems myself.

The experiment used a quantum dot photoluminescent source [5, 6] which was excited periodically by a stimulating oscillator and excitation laser at 75 MHz. During each period of the wave the dot emitted three photons as shown in figure 10.2. The correlations between the signals at the two detectors are shown graphically in figure 10.3. The x and y axes of the figure are time in ns during each period of the stimulation, so that you can see the correlations between each pair of

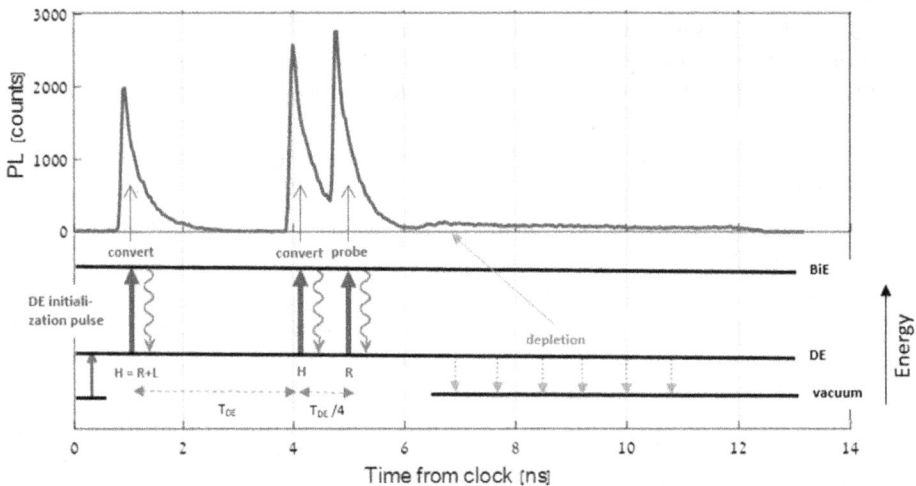

Figure 10.2. Light pulses emitted during a single period of the stimulating oscillator. Reprinted with permission from [4], Courtesy of G Peniakov.

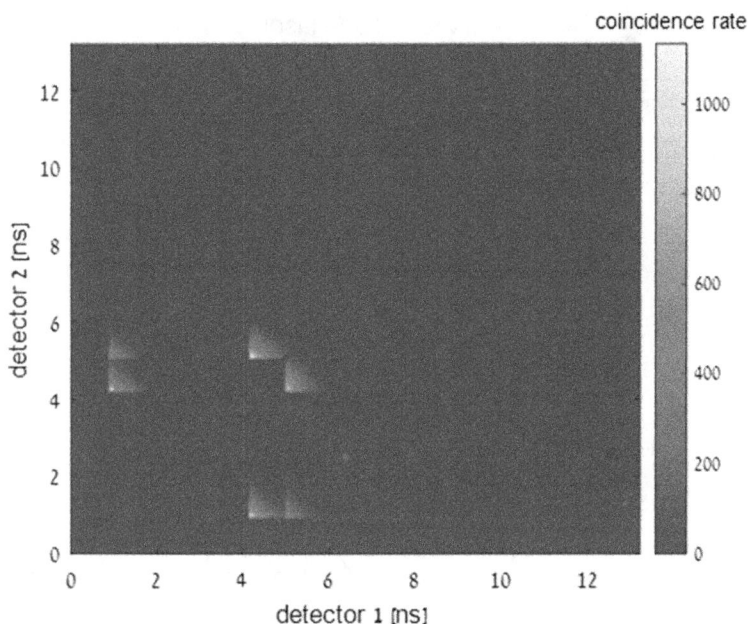

Figure 10.3. Correlations between the signals detected by the two detectors at the times indicated on the x and y axes during a single period of the excitation. The diagonal line at 45° through the origin indicates simultaneous detection, and no correlations are observed. Reprinted with permission from [4], Courtesy of G Peniakov.

emissions. However, what is important is that there are *no correlations* along the diagonal line at 45° which would represent the correlation between photons arriving at the two detectors simultaneously resulting from a single emission from the source. This cannot be explained using the wave description and is clear evidence for the quantization of the light field.

10.2 Measuring the Planck constant

There are several methods of measuring Planck's constant, and I will describe one optical method which is very simple and in my experience gives an excellent result [7]. Essentially, one measures the turn-on voltage of a semiconductor diode laser and the wavelength of its emission. The turn-on voltage can be found by using a variable voltage supply connected to the laser through an ammeter, and a voltmeter attached directly to the laser terminals. The light emitted from the laser can fall on a detector, or can be observed in the dark when the laser is directed onto a white scattering surface. When the voltage is raised, the current is at first very small, but undergoes a jump coincident with the first observation of light output. The current then increases until it reaches saturation at some higher voltage, which is the usual working voltage of the laser (figure 10.4). In a separate experiment, the laser wavelength should be determined using a diffraction grating spectrometer.

I(V) for diode laser

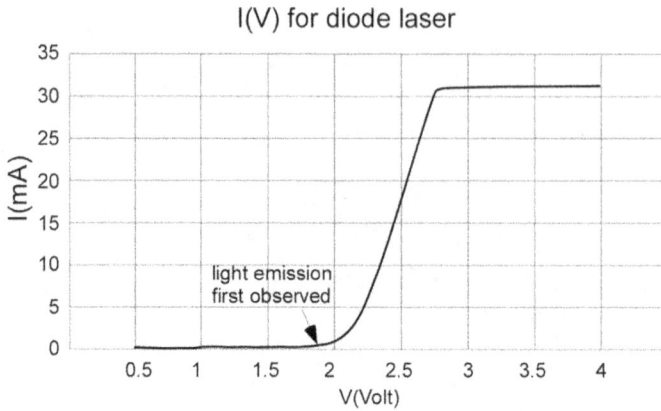

Figure 10.4. $I(V)$ for a laser diode, which emits at $\lambda = 639$ nm.

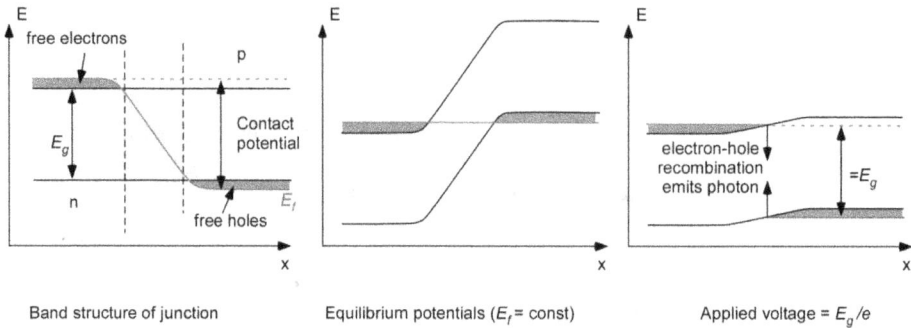

Band structure of junction   Equilibrium potentials (E_f = const)   Applied voltage = E_g /e

Figure 10.5. Application of voltage E_g/e results in photon emission.

A semiconductor diode laser in its basic form is constructed from a junction between heavily-doped p- and n-type direct band-gap semiconductors in which the recombination of a hole and an electron in the junction region results in emission of a photon [8]. Most direct band-gap semiconductors are of the III–V type, typically GaAs or its alloys with other tri- and pentavalent atoms. The first spontaneous light emission from the laser occurs when the voltage equals the semiconductor bandgap in electron volts (figure 10.5, right). As the voltage is increased, stimulated emission begins; the current grows significantly and eventually saturates and the optical bandwidth of the laser output narrows significantly. Modern heterostructure diode lasers are more complicated in that they may have three or more layers, the outermost being heavily-doped p- and n-type semiconductors and the light-emitting central layer being a pure semiconductor with a smaller bandgap than the outer layers, but the basic idea is the same. All this forms the basis of an experiment in which the current–voltage characteristic of the diode laser is carefully measured, together with the output spectrum of the laser light, using a diffraction-grating spectrometer. One can then measure the emission wavelength and deduce Planck's constant, and also observe the way in which the bandwidth narrows as the current increases and stimulated emission takes over.

10.3 Laser modes

The same laser as was used for determining Planck's constant can also be used to illustrate the idea of laser modes. When we look at the emission spectrum of the diode laser when the current is increased through the stimulated-emission threshold, we see the following. Below threshold, spontaneous emission is observed with quite a wide bandwidth. Then, a narrow stimulated emission line appears and becomes dominant (figure 10.6(a) i–iv). The wavelength of the stimulated emission is basically that for which the gain is maximum, which is determined by the optical cavity of the laser. The cavity of the laser is formed by its two parallel end-surfaces, separated by distance L, so that the allowed free-space wavelengths are the resonant wavelengths $\lambda = 2Ln/m$ where n is the refractive index and m is an integer. Each value of m defines an emission mode. Although one value of $m = m_0$ corresponds to the maximum gain, other wavelengths with close-to-optimal values of m are also emitted, and a banded spectrum is observed (figure 10.6(b)). From the mode

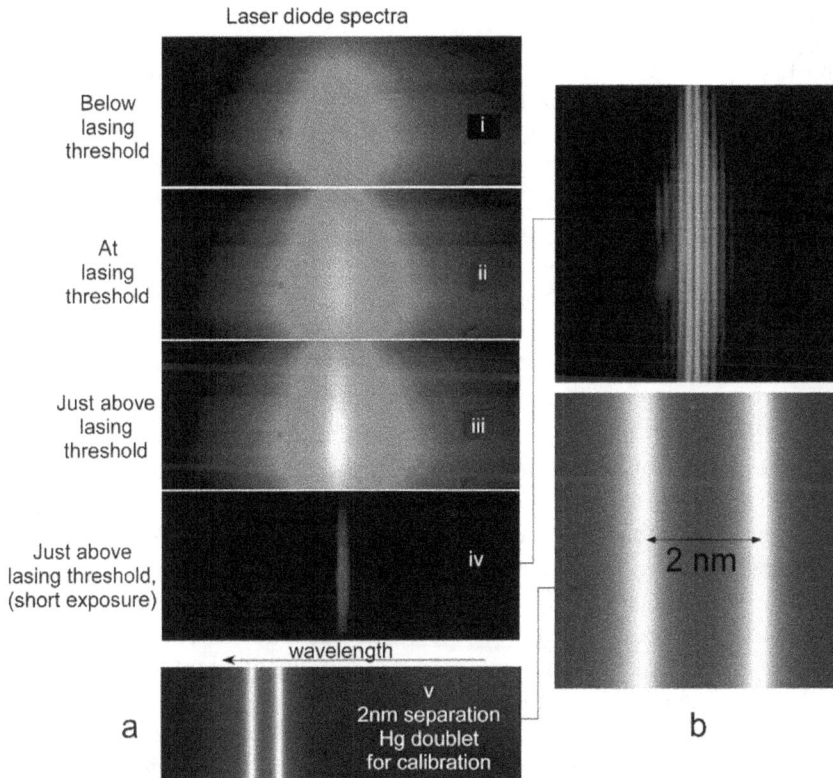

Figure 10.6. (a) Spectra recorded for a diode laser going through the lasing threshold, (b) Lasing mode spectrum. Notice the colour difference between the two reference Hg lines, separated by only 2 nm! The eye is very sensitive to colour differences in this region of the spectrum.

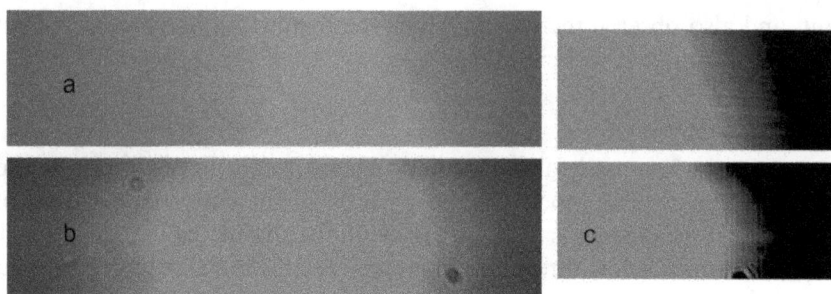

Figure 10.7. Comparison between the spontaneous emission spectrum below threshold (a) before lasing and (b) after lasing. (c) Note the residual mode structure (contrast enhanced) which can be seen in the lower spectrum, but is hardly visible in the upper one.

separation, the dimensions of the laser cavity can be deduced. It is interesting to note that when the laser current is reduced to below the threshold, so that only spontaneous emission is observed again, traces of the banded spectrum remain (figure 10.7).

10.4 The spectrum of black-body radiation

It is well-known that at the end of the 19th century, when it was thought that physics was essentially a closed book and that the future would only bring out more details in topics which were basically understood, one of the few still-outstanding problems was that of the spectrum of black-body radiation. This had been measured by William Herschel in 1800 [9], and showed that the radiation spectrum has a peak at a wavelength which depended on the temperature of the body, but there was no explanation for the fact. Indeed, the only existing classical theories of the spectral content of radiation in a closed cavity showed that the density of radiation per unit wavelength interval would increase at short wavelengths like $1/\lambda^4$ as $\lambda \to 0$, since each electromagnetic mode of the cavity would have zero-point energy of $k_B T$, where k_B is Boltzmann's constant. This was called the 'ultra-violet catastrophe'. Its solution by Planck led to the invention of quantum mechanics.

There are several publications of experiments at the student level designed to measure the spectrum of black-body radiation, which use infra-red spectrometers [10, 11]. The main problem with these experiments is that the instruments used may have been calibrated by using black-body radiation, which defeats their logic. One such point is that the temperature of the source used in the experiments, usually a tungsten filament lamp, is related to its electrical resistance, and the relationship may have been determined using a pyrometer to measure the filament temperature. Then, infra-red spectrometers are most usually calibrated in terms of black-body radiation. I therefore tried to devise an experiment which does not have these reservations in order to investigate the spectrum of a black body in the visible region quantitatively.

The following method seems to be feasible, but needs considerable additional work to improve its accuracy. It is based on the assumption that a CCD monochrome camera has quantum efficiency which is independent of wavelength, provided that this is shorter than the cutoff wavelength (of order 1 μm). I used a prism spectrometer (section 2.1), employing a low-dispersion BK7 glass prism with apex angle 30°, so that the whole visible spectral intensity could be photographed in one frame. To demonstrate the method, the wavelength scale of the spectrum was first calibrated by using a filament lamp source and a series of narrow-band filters at different known visible wavelengths. In principle these could have been independently checked by means of a diffraction grating spectrometer, or else a discharge lamp emitting a number of known spectral lines could have been used. Since the prism dispersion is not a linear function of the wavelength, the calibration results (wavelength as a function of position on the camera chip) were converted to an empirical function in order to make the observations quantitative.

I then used the Sun as a light source. We know that the Sun's radiation at ground level is not that of a black body, since it has passed through the atmosphere but it has peak intensity near 550 nm wavelength, which corresponds to a Planck function at 5800 K. It has a narrower band than a true black-body because of absorption in the ultra-violet and infra-red. The results are illustrated in figure 10.8. The peak was clearly visible, but at a wavelength of 600 nm. At first this was a disappointment, but then I realized that it was proof of the quantum nature of the radiation; since the photons at shorter wavelength have larger energy quanta, their number (for a given intensity) is smaller and this pushes the peak measured by a photon detector to a longer wavelength. In fact, using the Planck formula for photon flux does indeed move the Sunlight peak wavelength to approximately 600 nm, as can be seen in figure 10.9. Because of the atmospheric problem, the measured spectral intensity curve is not close to the Planck function, but this was expected. Although the experimental details need refinement (in particular, an independent confirmation

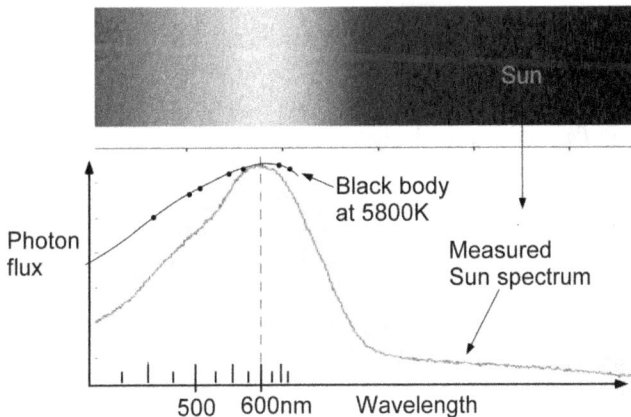

Figure 10.8. Comparison between the measured Sun photon flux spectrum and the corresponding Planck function for the Sun's surface temperature. The wavelength calibration is valid only in the region of the seven available narrow-band filters (shown as points on the black-body curve).

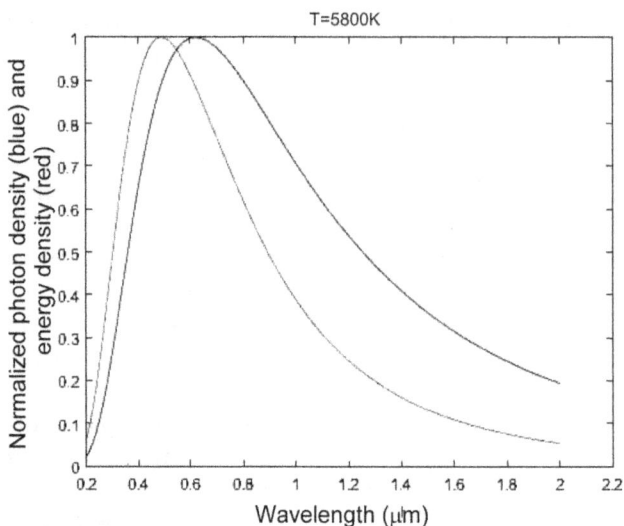

Figure 10.9. Planck black-body radiation curves at 5800 K, the Sun's surface temperature, showing the photon density (blue) and the energy density (red). The photon density peaks at 610nm, and the energy density at 500 nm wavelength.

that the quoted Sun surface temperature can be deduced independently from spectral line broadening or another phenomenon), I believe that since this experiment does not depend on any instrumental calibrations which may be traceable to the Planck function, it fulfils the requirement of an independent experimental proof of quantization of the electromagnetic field.

References

[1] Clauser J F 1974 Experimental distinction between the quantum and classical field-theoretic predictions for the photoelectric effects *Phys. Rev.* D **9** 853

[2] Aspect A, Grangier P and Roger G 1981 Experimental tests of realistic local theories via Bell's Theorem *Phys. Rev. Lett.* **47** 460

[3] Aspect A, Grangier P and Roger G 1989 Dualité onde-particule pour un photon *J. Opt.* **20** 119–29

[4] Peniakov G, Su Z-E, Beck A, Amar O, Ryzhkov M and Gershoni D 2019 Supersensitive optical phase measurements using deterministically generated multi-photon entangled states *Rochester Conf. on Coherence and Quantum Optics (CQO-11)(4–8 August 2019) (Rochester, New York, United States) OSA Technical Digest* (Washington, DC: Optical Society of America) paper M5A.17

[5] Schwartz I, Cogan D, Schmidgall E R, Don Y, Gantz L, Kenneth O, Lindner N H and Gershoni D 2016 Deterministic generation of a cluster state of entangled photons *Science* **354** 434–7

[6] Cogan D, Kenneth O, Lindner N H, Peniakov G, Hopfmann C, Dalacu D, Poole P J, Hawrylak P and Gershoni D 2018 Depolarization of electronic spin qubits confined in semiconductor quantum dots *Phys. Rev.* X **8** 041050

[7] Checchetti A and Fantini A 2015 Experimental determination of Planck's constant using light emitting diodes (LEDs) and photoelectric effect *World J. Chem. Educ.* **3** 87–92

[8] Siegman A E 1971 *Introduction to Lasers and Masers* (New York: McGraw Hill)

[9] White J R 2012 Herschel and the puzzle of infra-red *American Scientist* **100** 218

[10] Dryzek J and Ruebenbauer K 1992 Planck's constant determination from black-body radiation *Am. J. Phys.* **60** 251

[11] Severn R 2019 http://home.sandiego.edu/~severn/p272/blackbody_radiation_fa19.html

www.ingramcontent.com/pod-product-compliance
Lightning Source LLC
Chambersburg PA
CBHW080538220326
41599CB00032B/6306